The Minerals, Metals & Materials Series

Bowen Li · Baojun Zhao · Jian Li ·
Sergio Neves Monteiro · Zhiwei Peng ·
Dean Gregurek · Tao Jiang · Yong Shi ·
Cuiping Huang · Shadia Ikhmayies
Editors

Materials Engineering—From Ideas to Practice: An EPD Symposium in Honor of Jiann-Yang Hwang

TMS

Springer

Editors
Bowen Li
Michigan Technological University
Houghton, MI, USA

Jian Li
CanmetMATERIALS
Hamilton, ON, Canada

Zhiwei Peng
Central South University
Changsha, China

Tao Jiang
Central South University
Changsha, China

Cuiping Huang
Futianbao Environmental Protection
Technology Company
Xian, China

Baojun Zhao
The University of Queensland
Brisbane, Australia

Sergio Neves Monteiro
Military Institute of Engineering
Rio de Janeiro, Brazil

Dean Gregurek
RHI Magnesita
Leoben, Austria

Yong Shi
Futianbao Environmental Protection
Technology Company
Xian, China

Shadia Ikhmayies
Al Isra University
Amman, Jordan

ISSN 2367-1181 ISSN 2367-1696 (electronic)
The Minerals, Metals & Materials Series
ISBN 978-3-030-65240-1 ISBN 978-3-030-65241-8 (eBook)
https://doi.org/10.1007/978-3-030-65241-8

Cover illustration: From Chapter "Development and Management of an Industrial Park for the Chinese Electroplating Industry", Yong Shi et al., Figure 2: A part of the Futianbao wastewater treatment facility. https://doi.org/10.1007/978-3-030-65241-8_20.

This Springer imprint is published by the registered company Springer Nature Switzerland AG
The registered company address is: Gewerbestrasse 11, 6330 Cham, Switzerland

Preface

Professor Jiann-Yang Hwang is an internationally well-known expert with a career of over 40 years in the fields of mineral processing, metallurgy, water treatment, microwave-assisted material process, hydrogen storage, and by-product recycling. He also has more than 30 years of continuous service to the communities of The Minerals, Metals & Materials Society (TMS), the Society for Mining, Metallurgy, and Exploration (SME), American Chemistry Society (ACS), and China Non-ferrous Metal Society. His pioneering research achievements and leadership have already broadly influenced the society and industry internationally and have inspired young generations. To honor Professor Hwang's outstanding achievements and social services, TMS hosts a special symposium "Materials Engineering—From Ideas to Practice: An EPD Symposium in Honor of Jiann-Yang Hwang" during TMS 2021 Virtual Annual Meeting & Exhibition.

This symposium received 62 abstracts from 9 countries. The topics cover the broad area of minerals, metals, and polymers from material design, processing, characterization, to industrial applications, recycling, and environment protection, which mostly reflect Professor Hwang's research tracks and influence.

This proceedings volume collected 29 peer-reviewed manuscripts that will be presented in the symposium, most of which are invited presentations. It focuses on the novel science and technology progress in characterization and processing development in the fields of minerals, metals, and materials, and multidisciplinary interactions. This book provides inspiration to the readers in academia and the manufacturing industry.

This symposium is sponsored by the TMS Extraction and Processing Division and the Materials Characterization Committee and Pyrometallurgy Committee. The organizing team appreciates all the authors, presenters, and editors for their support and help.

Bowen Li
Lead Organizer

Contents

About the Honoree

Jiann-Yang Hwang is honored for pioneering research on microwave-assisted metallurgy; for engineering and commercialization of environmental technologies to the electroplating industries; for outstanding innovations on mineral processing; for crucial achievements to power, automobile, steel, and aluminum industries on materials recycling; for great contributions on hydrogen storage materials; for significant developments on clean coal technologies; for exceptional leadership and service to TMS; and for excellent student mentoring.

Jiann-Yang Hwang was born on December 4, 1952, in Taipei, Taiwan. His father, Mr. Wei Hwang from Daye, Hubei, China, was an Army General. His mother, Siping Tang from Zhijiang, Hunan, China, was a school teacher. Correct pronunciation of Jiann-Yang Hwang is like "Jan Young Whang". He is called by his friends "Jim Hwang" most of the time.

He received his B.S. degree in Earth Sciences from the National Cheng Kung University, Taiwan, in 1974. He moved to the United States in 1977 and obtained his M.S. degree in Earth and Atmospheric Sciences in 1980 and Ph.D. in Mineral Processing and Metallurgical Problems from Purdue University in 1982. After completing his postdoc research from the School of Electrical Engineering at Purdue with emphasis on electromagnetics in 1984, he joined Michigan Technological University and has stayed there until now. He has been the Director of the Institute of Materials Processing and Professor and Chair of the Mining Engineering Department, Tenured Professor of Materials Science and Engineering, and Professor and Adjunct Professor in Environmental Engineering, Chemical Engineering, Electrical Engineering, and Geology.

Over the past 40 years, Dr. Hwang has dedicated his life to research, education, and industrial consulting services. He is the author/co-author of over 200 research papers, editor/co-editor of 22 books, inventor/co-inventor of over 30 patents, and PI/Co-PI of more than 100 research projects funded by government agencies and industries. His research on microwave-assisted steel production received the award of "Grand Challenge on the Steelmaking Technology of the Next Generation" from the U.S. Department of Energy. His research on hydrogen storage materials also won the DOE "Grand Challenge on Hydrogen Storage Materials" award. His flyash beneficiation technology was licensed and utilized by the Mineral Resource Technology, Inc., a subsidiary of CEMEX, the largest cement and construction materials company of the world. Futianbao Environmental Protection Technologies Company has built and operated the "Zero Liquid Emission" commercial plants in China to treat the wastewater of various Cu, Ni, Zn, Cr and other electroplating industries. These are a few examples of his achievements.

Prof. Hwang received TMS EPD Technology Award (2011), Characterization Committee Best Paper Award (2012, 2016, 2019), Michigan Technological University Bhakta Rath Research Award (2013), China Nonferrous Metal Society First Class Paper Award (2017), AIME James Douglas Gold Medal (2019), and 1000 Talents Award of China (2011). He was engaged as the Chief Energy and Environmental Advisor by Wuhan Iron and Steel Co. Group (2011–2015), a Fortune 500 company, and the Chief Scientist of Futianbao Environmental Protection Technology Company (2015–present). He was also awarded with the Honorary Guest Professor by Beijing University of Science and Technology, Kunming University of Science and Technology, Central South University, North China University of Science and Technology and Chongqing University, and has also been invited to lecture in the Indian Institute of Technology, Kanpur.

Prof. Hwang also exhibited his exceptional leadership in professional society services. He joined TMS and SME in the 1980s and has continued his service through today. He served as the Chairperson for TMS technical committees including the Process Mineralogy, Characterization, and Pyrometallurgy for several times. He is the founder, organizer, and co-organizer of TMS Characterization of Minerals, Metals and Materials Symposium (2004–present) and the International High Temperature Processing Symposium (2009–present). Besides that, he organized/co-organized several other symposia for TMS meetings and chaired a number of technical sessions. With his great efforts and leadership, both symposia became the annual primary symposia of the Characterization Committee and Pyrometallurgy Committee. The series of proceedings of "Characterization of Minerals, Metals, and Materials" and "International High Temperature Processing" are two major proceedings of the TMS Annual Meeting. In 2019, he received the TMS EPD Distinguished Service Award in honor of his long-term leadership in the TMS community.

In addition, Prof. Hwang was an active leader in mineral processing, metallurgy, water treatment, and waste materials reuses in the Society for Mining, Metallurgy, and Exploration (SME), and American Chemistry Society (ACS), as a symposium organizer and session chair over the past years. He also served as the Editor-in-Chief for the *Journal of Mineral and Materials Characterization and Engineering* (2001–present), Editor of *Current Microwave Chemistry* (2013–present), and Editor of *International Journal of Mineral Processing* (1990–2000).

Dr. Hwang has been dedicated to undergraduate, graduate, doctoral, and postdoctoral education. Examples of some worthy mentoring activities include:

(i) Rath Research Award of Michigan Technological University, 2013. This award recognizes a Ph.D. student and the advisor for their exceptional scientific and technological research in anticipation of the future needs while supporting potential advances in emerging technology.
(ii) First Place, MTU Senior Design annual competition of the whole campus, 2008 (advised a Materials, Mechanical, and Business enterprise team).
(iii) Championship, Graduate Student Team, Automotive Solutions Competition on Plastics Recycling, 1994, jointly sponsored by American Plastics Council, Vehicle Recycling Partnership (the Big Three: GM, Ford, and Chrysler) and Society of Automotive Engineers.
(iv) TMS Young Professional Leader Award to one of his Ph.D. students.

We are very happy to have this symposium honoring Professor Jiann-Yang Hwang. His contributions and achievements to the science, technology, society, and education are very appreciated.

About the Editors

Bowen Li is a research professor in the Department of Materials Science and Engineering and Institute of Materials Processing at Michigan Technological University. His research interests include materials characterization and analysis, metals extraction, ceramic process, antimicrobial additives and surface treatment, porous materials, applied mineralogy, and solid waste reuse. He has published more than 120 technical papers in peer-reviewed journals and conference proceedings, authored/co-authored 3 books, editor/co-edited 12 books, and holds 15 patents.

Dr. Li received a Ph.D. degree in Mineralogy and Petrology from the China University of Geosciences Beijing in 1998, and a Ph.D. degree in Materials Science and Engineering from Michigan Technological University in 2008. He has been an active member of The Minerals, Metals & Materials Society (TMS), Society for Mining, Metallurgy, and Exploration (SME), and China Ceramic Society. At TMS, he currently serves as the past chair of the Materials Characterization Committee and a member of the Powder Materials Committee and Biomaterials Committee, and formerly as EPD Award Committee member, *JOM* subject advisor, and key reader of *Metallurgical and Materials Transactions A.* He is the organizer/co-organizer of a number of international symposia and sessions. He also served as an editorial board member of the *Journal of Minerals and Materials Characterization and Engineering*, *Reviews on Advanced Materials Science*, and *FUTO Journal Series.*

Baojun Zhao is a professor in the College of Materials, Metallurgy, and Chemistry at Jiangxi University of Science and Technology, Ganzhou, China. His primary fields of research are fundamental and applied investigations relevant to high-temperature processing of metals and materials. He has developed a series of novel research techniques to enable high-quality research to be carried out. He has published over 200 refereed journal and conference papers and received a number of international awards to demonstrate his leading research achievements. He has long-term collaborations with many international companies on metallurgy and resources including Baosteel, Codelco, Dongying Fangyuan Nonferrous Metals, HBIS, Pangang Group, Rio Tinto, and Shougang.

Jian Li is a senior research scientist and program manager at CanmetMATERIALS in Natural Resources Canada. He obtained his B.Sc. in Mechanical Engineering from Beijing Polytechnique University, M.Sc. in Metallurgical Engineering from Technical University of Nova Scotia, and Ph.D. in Materials and Metallurgical Engineering from Queen's University, Kingston, Ontario, Canada. He has broad experience in materials processing and characterization including alloys deformation, recrystallization, and micro-texture development. Dr. Li has extensive experience in Focused Ion Beam (FIB) microscope techniques. He is also an expert in various aspects of SEM-EDS and EPMA techniques. Dr. Li has authored three book chapters and published more than 160 papers in scientific journals and conference proceedings.

Sergio Neves Monteiro graduated as a metallurgical engineer (1966) at the Federal University of Rio de Janeiro (UFRJ). He received his M.Sc. (1967) and Ph.D. (1972) from the University of Florida, followed by a 1975 course in energy at the Brazilian War College, and a postdoctorate (1976) at the University of Stuttgart. In 1968 he joined the Metallurgy Department of UFRJ as a full professor of the postgraduation program in engineering (COPPE). He was elected the head of department (1978), coordinator of COPPE (1982), Under-Rector for Research (1983), and was invited as the Under-Secretary of Science for the State of Rio de Janeiro (1985) and Under-Secretary of the College Education for the Federal Government (1989). He retired in 1993 from the UFRJ and joined the State University of North Rio de Janeiro (UENF), where he retired in 2012. He is now a professor at the Military Institute of Engineering (IME), Rio de Janeiro. He has published more than 1,500 articles in journals and conference proceedings and has been honored with several awards including the ASM Fellowship and several TMS awards. He is the top researcher (1A) of the Brazilian Council for Scientific and Technological Development (CNPq) and top scientist of the State of Rio de Janeiro (FAPERJ). He was the president of the Superior Council of the State of Rio de Janeiro Research Foundation, FAPERJ (2012), and currently is the coordinator of the Engineering Area of this foundation. He has also served as the president of the Brazilian Association for Metallurgy, Materials, and Mining (ABM, 2017–2019), as a consultant for the main Brazilian R&D agencies, and as a member of the editorial board of five international journals as well as associate editor-in-chief of the *Journal of Materials Research and Technology*. He is the author of 30 patents and a top world researcher in "Natural Fiber Composites" and "Ballistic Armor", Scopus 2029.

Zhiwei Peng is a professor in the School of Minerals Processing and Bioengineering at Central South University. He received his B.E. and M.S. degrees from Central South University in 2005 and 2008, respectively, and his Ph.D. degree in Materials Science and Engineering from Michigan Technological University in 2012. His research interests include heat transfer in microwave heating, dielectric characterization of materials, nonthermal microwave effects, extractive metallurgy, computational electromagnetics, microwave absorbing materials, and biomaterials.

He has published 140 papers, including more than 100 peer-reviewed articles in journals such as *International Materials Reviews*; *Journal of Hazardous Materials*; *ACS Sustainable Chemistry & Engineering*; *Resources, Conservation & Recycling*; *Journal of Cleaner Production*; *Metallurgical and Materials Transactions A*; *Metallurgical and Materials Transactions B*; *JOM*; *Journal of Power Sources*; *Fuel*

Processing Technology; *Energy & Fuels*; *IEEE Transactions on Magnetics*; *IEEE Transactions on Instrumentation and Measurement*; *Ceramics International*; *Powder Technology*; and*Separation and Purification Technology*. He has served as an associate editor for *Mining, Metallurgy & Exploration* since 2019, as a guest editor for *JOM* since 2013, and as an editor for *PLOS ONE* and *Cogent Chemistry* since 2018. He has been a member of the editorial boards of *Journal of Minerals and Materials Characterization and Engineering* since 2012, *Scientific Reports* since 2019, and *Journal of Central South University* since 2020, and has served as a reviewer for more than 70 journals. He received a TMS Travel Grant Award for the 141st TMS Annual Meeting & Exhibition, the Doctoral Finishing Fellowship and Dean's Award for Outstanding Scholarship of Michigan Technological University in 2012, and the Bhakta Rath Research Award of Michigan Technological University in 2013.

Dr. Peng is an active member of The Minerals, Metals & Materials Society (TMS). He has co-organized 9 TMS symposia (Characterization of Minerals, Metals, and Materials in 2013–2018, and the 9th–11th International Symposia on High-Temperature Metallurgical Processing in 2018–2020) and co-chaired 20 symposia sessions since 2012. He is a member of the Pyrometallurgy and Materials Characterization Committees and was the chair of the Continuing Education Sub-Committee of the Materials Characterization Committee. He was a winner of the TMS EPD Young Leaders Professional Development Award in 2014.

Dean Gregurek is a senior mineralogist in the RHI Magnesita Technology Center Leoben, Austria since 2001. He received his M.Sc. degree at the University of Graz in 1995 and his doctorate degree in Applied Mineralogy from the University of Leoben in 1999. Prior to RHI Magnesita, he worked for 2 years for Luzenac Europe in the talc business. His current research interests and technical expertise are focused on chemical and mineralogical studies related to interactions between refractories, molten metals, and slags from pyrometallurgical furnaces. He has been a TMS member since 2012, *JOM* advisor (2014–2017), chair of the Pyrometallurgy Committee (2018–2020), and a co-organizer for the 7th–11th International Symposium on High-Temperature Metallurgical Processing (TMS Annual Meetings 2016–2020).

Tao Jiang received his M.S. in 1986 and Ph.D. in 1990, both from Central South University of Technology. Then he joined the university and served as an assistant professor (1990–1992) and full professor (1992–2000). From 2000 to 2003, he was a visiting scientist to the Department of Metallurgical Engineering, the University of Utah. Since 2003, Dr. Jiang has been a Professor in the School of Minerals Processing & Bioengineering at Central South University. He has been Specially Appointed Professor of Chang Jiang Scholar Program of China since 2008 and dean of the school since 2010.

His research interests include agglomeration and direct reduction of iron ores, and extraction of refractory gold ores. He has undertaken more than 50 projects from the government and industry, including the National Science Fund for Distinguished Young Scholars Program. He and co-workers invented the direct reduction process of composite binder pellets and three plants were set up in China based on the invention. He proposed the innovative composite agglomeration process of iron ore fines, which was put into production in Baotou Steel Company, China. He is actively involved in the areas of utilization of non-traditional ferrous resources such as complex ores and various solid wastes. He has published 320 technical papers, 6 books including *Direct Reduction of Composite Binder Pellets and Use of DRI*, *Principle & Technology of Agglomeration of Iron Ores*, *Chemistry of Extractive Metallurgy of Gold*, and *Electrochemistry and Technology of Catalytical Leaching of Gold*. He holds 35 patents and has more than 30 conference presentations.

Yong Shi joined Futianbao Environmental Protection Technology Company in 2008, which is a business focused on electroplating wastewater treatment. He is the CEO of the company and has managed to make the company grow from a regional business in Xi'an to a national business with expansions of its business into Jiangsu, Zhejiang, and other provinces in China. He has cooperated with the Honoree, Dr. Jiann-Yang Hwang, and put significant resources on research and development of electroplating technology. The results are very fruitful. Futianbao is the leading company in China that achieved the Zero Liquid Emission status for electroplating enterprises. Various metallurgical and environmental technologies have been developed and utilized by the company to separate and purify metals, water, and salts from electroplating wastewater and get them recycled. Their technologies have received a special award for distinguished contributions and are promoted to the electroplating industries by the Chinese Environmental Protection Association.

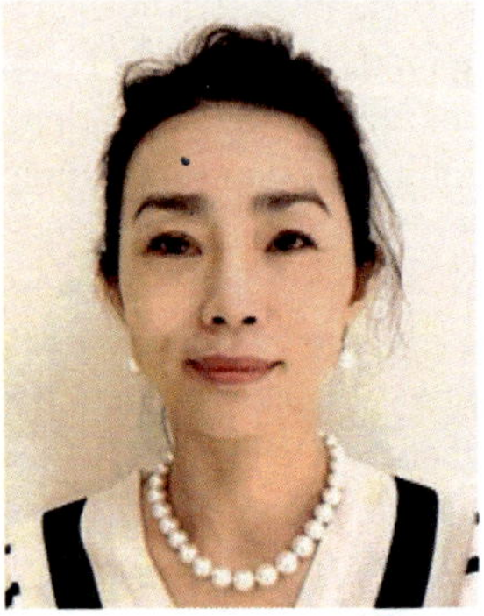

Cuiping Huang graduated from Xi'an Jiao Tong University in 1995 with a major in Business Administration with an emphasis on Economics. She is the president of the Futianbao Environmental Protection Technology Company, which she founded in 2000. Futianbao has its business focused on electroplating wastewater treatment. Her company invented the first heavy metal catching agents for the electroplating wastewater treatment, which also find a use for other heavy metal-bearing water treatment, such as mine water. The agent precipitates heavy metals through chelating reactions and reduces the heavy metal concentration from electroplating wastewater to an extremely low level. Futianbao is also the first company with the vision of electroplating industrial parks that place various kinds of electroplating enterprises altogether. This allows the electroplating companies to focus on their business and let the experts handle their environmental management and winning trust from the government agencies. The company built and operated the first electroplating industrial park in Northwest China. Its Xi'an park was completed in 2003, which has 200 U.S. acres and houses 40 electroplating companies. All the wastewater from these electroplating enterprises are collected and treated by Futianbao. Many technologies employed by the company are presented in this symposium.

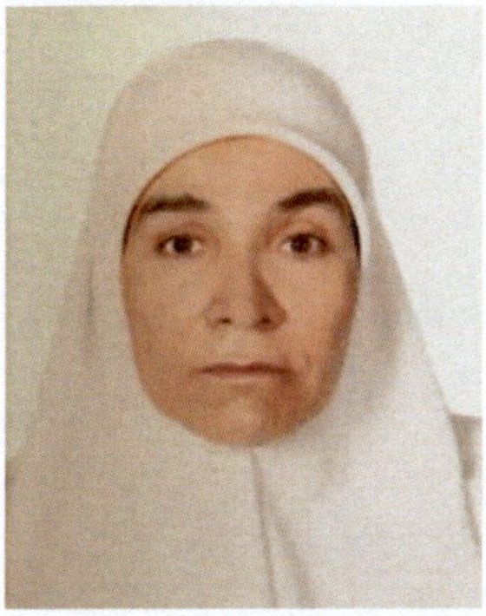

Shadia Ikhmayies received B.Sc. and M.Sc. degrees from the physics department at the University of Jordan in 1983 and 1987, respectively, and a Ph.D. in producing CdS/CdTe thin-film solar cells from the same university in 2002. Her research is focused on producing and characterizing semiconductor thin films and thin-film CdS/CdTe solar cells. She also works in characterizing quartz in Jordan for the extraction of silicon for solar cells and characterizing different materials by computation. She has published 56 research papers in international scientific journals, 80 research papers in conference proceedings, and 3 chapters in books. She is the founder and editor of the eBook series *Advances in Material Research and Technology* published by Springer, and the editor-in-chief/editor of several books.

She is a member of The Minerals, Metals & Materials Society (TMS) where she was the chair of the Materials Characterization Committee (2016–2017), and the leading organizer of three symposia: Solar Cell Silicon 2017–2020, Mechanical Characteristics and Application Properties of Metals and Non-metals for Technology: An EPD Symposium in Honor of Donato Firrao, and Green Materials Engineering: An EPD Symposium in Honor of Sergio Monteiro. She is also a member of the World Renewable Energy Network/Congress (WREN/WREC) 2010–present. She is a member of the international organizing committee and the international scientific committee in the European Conference on Renewable Energy Systems (ECRES2015–ECRES2020). She is a guest editor and a member of the editorial board of several journals including *JOM* and the *Journal of Electronic Materials*. She is a reviewer of 24 international journals and several international conference proceedings.

Part I
Mineral and Material Processing

Materials Processing, from Ideas to Practice

Jiann-Yang Hwang

Abstract Dr. Jiann-Yang Hwang served as the Director of the Institute of Materials Processing at Michigan Technological University for more than 20 years. Many technologies have been developed from ideas to commercial practices in his career life. This symposium reflects his contributions in this aspect. The cycle of materials such as metals on the earth involved the steps of ore exploration (geology), mining, mineral processing, metallurgy, manufacturing, and recycling. Each step is achieved by processing materials using energy. Depending on the process and the forms of energy input, products, and by-products with various environmental impacts are generated through air, water, and solid means. To obtain the most efficient process with the minimum environmental impacts at the best economics is the driving force that continuously pushes the advances of technologies. Variables in the materials, process, and energy are common parameters facilitating the development of ideas for technology advancements. Dr. Hwang learned earth sciences, mineralogy, characterization, mineral processing, and metallurgy during his undergraduate and graduate studies. Mining, materials, and processing, and environmental and economics are mostly self-studied at postgraduate time, partly pushed by the research needs from projects he wanted to conduct. Understanding the parameters involved in the materials, energy, environment, and economics is fundamental to a systematic approach. The validity of ideas and their potential to move to practice depend on the soundness of the system. The author reviewed several cases of his research to illustrate their relations.

Keywords Mineral processing · Magnetic separation · Microwave metallurgy · Recycling · Hydrogen storage · Water treatment · Material lifecycle

---

J.-Y. Hwang (✉)
Michigan Technological University, Houghton, MI 49931, USA
e-mail: jhwang@mtu.edu

B. Li et al. (eds.), *Materials Engineering—From Ideas to Practice: An EPD Symposium in Honor of Jiann-Yang Hwang*, The Minerals, Metals & Materials Series,
https://doi.org/10.1007/978-3-030-65241-8_1

Introduction

I am deeply touched to have this symposium sponsored by TMS that honors my lifelong contributions to the society, the education, and the industry. This is a big milestone for me. This is also the time to look back, to find out what has been important to shape me, what I have contributed, and what are the experiences I can share with others.

There are many people and organizations that I appreciate. My family, my teachers, my colleagues, my students, my friends, and even those people who built barriers to me, I want to say "thank you". I have learned a lot from each of you. I will try to tell a few stories here. Some of your names will be mentioned. Most will not. This does not mean you are less important to my life. I am simply not able to name each of you, but you definitely left marks in my heart, many are deeper than those I mentioned.

Life is a continuous learning process. I am so fortunate that new things keep on hitting me that provided me the opportunity of continuous learning. The funding agencies always challenge me with new problems. The industry and friends keep on asking me if I can help finding solutions.

Teamwork is very important. Nobody knows everything. We always need assistance from others. Colleagues, friends, and students are never shy to provide their two cents and work together. The critical thing is how much one can absorb and put things together. Continuous learning is so important here. Only the person who is willing to think, to learn from the information on various aspects, to learn from others, and to put things together will be able to lead and be successful.

What Has Been Done

I was born in 1952 in Taiwan. My father was an Army General and I was the fourth son with a younger sister. His interest was to put all of us in the military since his life was one war after another. However, none of us obeyed that. My name is Jiann-Yang. All of my brothers has the "Yang" part, which is our generation name. The "Jiann" part means "build". That was my excuse not to go to the military school since that is not to build.

Like everybody in elementary school, my fifth-grade teacher asked us to write an essay on what we would like to be after we grew up. My answer was I want to be an inventor like Edison. I want to "Build" things, especially new things that can help society.

Nobody teaches how to be an inventor at my time. My only idea was to study science and engineering. My father's hometown, Daye, which means Big Smelting, was famous for its iron and copper ores and smelting. So, my first college major was Earth Science at the National Cheng Kung University. I received my B.S. degree as a top student there in 1974. After 2 years of military service as a second lieutenant in the army, I joined the Mining Research Institute in Taiwan as a mineral exploration engineer with a focus on remote sensing. In 1977, I obtained a scholarship and came to the United States to study mineralogy and geochemistry in the Earth and Atmospheric Science Department at Purdue University. Scanning electron microscope and electron microprobe analysis was new at that time and I was lucky to be one among the early generation who can look into the microstructure of materials. After completing my M.S. in 1980, I obtained support from the Electric Engineering School and began my PhD study in the mineral processing field.

I never thought I could become an inventor until my Ph.D. study at Purdue University. My Ph.D. advisor, Professor Fritz Friedlaender, was the President of the IEEE society in 1977 with a specialty in magnetics. In addition to the magnetic bubble memory area, he was also a pioneer in the High Gradient Magnetic Separation field. That was just developed to separate weakly paramagnetic material from diamagnetic materials at an ultrafine particle size and applied in the mineral industry such as clay, uranium, and rare earth. I learned Electromagnetics and Separation from him which assisted me making many unconventional breakthroughs in various fields.

I was also advised by Prof. Reinhardt Schuhmann, Jr., from Chemical and Metallurgical Engineering. Prof. Schuhmann was a colleague of Prof. Antoine Gaudin at the Massachusetts Institute of Technology in ore dressing (mineral processing) before he moved to Purdue. The famous Gaudin–Schuhmann Model on particle size distribution was given by them. I was also co-advised by Prof. Phillip Wanket in Chemical and Metallurgical Engineering and Prof. Gunnar Kullerud in Earth Science. Prof. Wanket is well known for his book "Separation Process Engineering: Includes Mass Transfer Analysis". Prof. Kullerud was famous in the sulfide mineralogy, who established many phase diagrams of sulfides and advised mining companies on their applications to mineral exploration. I build my knowledge in various areas from them.

My Ph.D. thesis was my first challenge on innovation. I need to separate alunite from quartz in a Colorado ore after grinding the ore to an average particle size of 5 μ when ultrafine particle processing technology was lacking or nonexisting. Alunite is a potassium aluminum sulfate mineral. It is diamagnetic (or non-magnetic) and so does the quartz. To separate them by magnetic separation is nearly impossible, but it has to be done. I could not graduate unless I solve the problem, as my advisor told me. I have to detour from conventional physics and make alunite pseudo-magnetic. I initiated the magnetic seeding concept to selectively co-flocculate alunite and magnetite since I found an article from Prof. Kitchener in Empire College of England talking about selective flocculation of ultrafine particles for their separation. If one can selectively flocculate a mineral, then it might be possible to flocculate two minerals together. After I learned and was able to control the surface chemistry of various minerals and their behavior in water, I made the magnetic separation of alunite and quartz [1, 2] and got my Ph.D. degree.

I was a postdoc in the School of Electric Engineering at Purdue University from 1982 to 1984 to assist Prof. Friedlaender in continuing the development of the theory and application of high gradient magnetic separation technology. In this period, I had my first patent, "Magnetic Separation Method Utilizing a Colloid of Magnetic Particles" (US Patent 4,526,681). My colleague, Dr. Makoto Takayasu, who moved to

MIT later, and I were playing ferrofluid or called magnetic fluid at that time. Ferrofluid was a new material found applications in seal, magnetic levitation, etc. It is composed of magnetite particles of about 100 angstroms homogeneously dispersed in water by lignosulfonate-type coating on the nanoparticles [3]. Most people use ferrofluid based on their bulk property. However, we found that the diffusion of the magnetic particles under the magnetic force can yield a magnetite concentration gradient. Then particles of different materials possessing different magnetic susceptibilities placed in the fluid with magnetite nanoparticle concentration gradient under a magnetic field can move to positions where their magnetic susceptibility equals to the magnetic susceptibility of the colloid with corresponding concentration. This is the first time one can separate materials purely based on their magnetic susceptibility without worrying about other physical properties such as particle size and density [4].

Magnetic fluid is composed of nanomagnetic material coated with charged organics such as lignosulfonate, a paper pulping by-product, to provide sufficient repulsive force between the nanomagnetic particles. Learned from that concept, I combined it with the flotation collector concept to create a new kind of mineral processing technology. Froth flotation is the most commonly utilized separation technology in mineral processing. This technology was invented and developed by Prof. Gaudin of MIT, a colleague of one of my advisor, Professor Reinhardt Schuhmann, Jr. Froth flotation uses the selective adsorption of surfactant on a mineral particle through the functional group of the surfactant (surfactant has one end of functional group with hydrophilic property and the other end of hydrocarbon chain with hydrophobic property) to render the mineral particle hydrophobic and hence make the particle possessing the ability to attach to an air bubble for its flotation so the separation of different minerals can be achieved. Through the years of development, the chemical system for the selective adsorption of different minerals has been established and applied in the industry. However, froth flotation has problems to apply to particles of microns and below. Ultrafine particles generally cannot be made hydrophobic enough in water to facilitate efficient particle–bubble attachment. For example, kaolin clay is a major ingredient for paper coating to make the paper white and easy to write. Kaolin clay has impurities such as anatase TiO_2 that makes clay not white enough and has to be removed after mined. Since the anatase particles in clay are microns and less, froth flotation was difficult to be applied and High Gradient Magnetic Separation was utilized by the clay industry. Anatase is only feebly magnetic, this gives a big headache to the clay industry. They have to increase the magnetic field intensity at very high cost to enhance the separability.

I invented a new technology ("Reagents for Magnetizing Nonmagnetic Materials", US Patent 4,834,898 (1989), 4,906,382 (1990), and 5,043,070 (1991)) that opens the door for the treatment of ultrafine particles. In this technology, nanomagnetic particles are coated with two layers of surfactants like a Michelle structure. The first (inner) layer has a functional group with an affinity to the magnetic particle. The second (outer) layer has an affinity to the particle we want to separate. Through this arrangement, nanomagnetic particles can be selectively adsorbed on the particles of interest [5, 6]. Cytec Industries, a billion-dollar revenue major specialty chemical company that has business in froth flotation chemicals has commercialized this technology.

I have applied the concept of high electric charge/polarization to enhance the adsorption of hydrogen for hydrogen storage materials [7–11]. My hydrogen storage material concept received one of the Grand Challenge awards of the US Department of Energy.

Electrosorption technology was a water treatment technology that has gone all the way to the full commercial practice [12]. In its operation, water flows in between the electrodes. Ions and charged molecules are attracted to the electrodes of opposite charge. Once they reach the electrodes, they are adsorbed and stored there. Thus, they are separated from clean water. When the adsorption is saturated, the circuit is shorted to release the ions and flush them out. In addition to various operational parameters of the system, such as the voltage, flow velocity, retention time, etc., electrode material is the key to the electrosorb technology. Porosity, pore structures, and chemical stability are some of the important factors.

The water treatment technology was further considered in a system that treats the electroplating wastewater using a variety of technologies including membranes and electrodialysis [13]. The electroplating wastewater is one of the most complicated industrial wastewaters that contains chemicals and wastes generated from various plating operations such as copper, nickel, chrome, zinc, and others. Many organic and inorganic materials are involved for the quality of the plating. Futianbao is a leading company in China for electroplating wastewater treatment and there are several papers in this symposium discussing the advances on recent developments.

Development of copper-based antimicrobial coating material that can inactivate bacterial and virus on the surface of products to avoid sickness through contact [14]. Its commercial development was carried out by Dr. Bowen Li under the QTEK Inc.

I have conducted research on materials recycling where I combine separation technology, materials processing technology, market analysis, and re-manufacturing and re-utilization technologies with a life cycle analysis to obtain the most logical approach. Fly ash, automobile shredding residue, steelmaking slags, foundry sands, grinding swarfs, blasting media, aluminum smelting slags and saltcakes, are a few examples of the projects he directed [15–21]. Cemex, one of the world largest cement and constructional material company with 18 billion in revenue, has licensed the flyash separation and utilization technologies I developed (US Patent 4,834,898 (1989); 4,906,382 (1990); 5,047,145 (1991); 5,249,688 (1992); 5,096,572 (1993); 5,277,047 (1993); 6,068,131 (2000)), which makes millions tons of coal combustion by-products generated from power plants to quality products that can be utilized in constructions, plastic fillers, and others.

One of the biggest challenges to me is the microwave metallurgy technology, which I have spent the last 20 years to advance the theory, mechanisms, equipment, and operations for its industrial applications [22–33]. The environmental problems are the biggest challenge to the smelting operations of all metal industries. When I was young, I observed the closure of one of the largest copper ore smelters in my neighborhood at White Pine, Michigan. In the 1990s, the bankruptcy of more than

30 of the world's largest and most famous iron and steel companies, such as Inland, Bethlehem, LTV, etc., astonished me and my generation. I began to think about what I can do in my life to change that, making the industry and environment more friendly to each other.

The fundamental problem is all the smelting technologies are basically combustion-oriented. Combustion inevitably resulted in various air pollutants such as NOx, SOx, CO_2, dust, etc., that are very burdensome to the smelting operations. To avoid combustion, air-free heating and reaction are required. Since carbohydrates, carbon, water, iron oxides, and many other minerals are microwave absorbing materials, I began to investigate the possibility of microwave metallurgy. It is easy to understand that CO_2 and H_2O react with carbon at high temperature to generate CO and H_2 gases (Boudouard reaction). CO and H_2 gases are excellent reductants to reduce metal oxides to metals and generate CO_2 and H_2O again. As long as carbon and microwave heating exist, the reaction will continue until all oxides reduced and the final products are only metal, CO, and H_2. In a sealed microwave furnace, no air is present so NOx is prevented. No blast air and hence no dust. Sulfur is converted mostly to CaS instead of SOx. The reaction can be on–off controlled, enabling the use of cheap energy available off the peak, such as nights. Electric energy is clean and can be generated from all kinds of technologies. Smelters will only need to pay without worrying about environmental emission punishing tickets and lawsuits. CO and H_2 are the Syngas components that can be the raw material for the production of all kinds of chemicals and fuels. Not only coal but biomass wastes can also a good carbohydrate source in this approach, solving another environmental problem.

Development of microwave metallurgy is a long and difficult road. The chemical reactions are easy to prove. Through even a household microwave, I have proved that metal can be produced using this new concept. An early patent, "Method for Direct Metal Making by Microwave Energy", US Patent 6,277,168, 2001, demonstrated that. However, it is far away from industrial practice. How to transport the microwave from its generator at a distance, how to design a furnace that can meet the needs to distribute the microwave to materials evenly and efficiently, what are the interactions between microwave and materials under various physical and chemical conditions, how to obtain the highest heating efficiency, what are the mass transfer and heat transfer phenomena and how to control them for the reaction uniformity, etc. There are numerous problems associated with it that are not traditionally encountered by metallurgists. Dr. Zhiwei Peng, Dr. Xiaodi, and I have tried to solve many fundamental and operational problems but much more work is still needed.

Career Development

After my postdoc research at Purdue from 1982 to 1984, I joined Michigan Technological University in 1984 and stayed there all the time. I started as a Research Scientist at the Institute of Mineral Research and had the adjunct assistant professor appointment in the Department of Electrical Engineering and the Department of

Geological Engineering in the university. The university was established in 1885 as the Michigan School of Mines because of the rich copper and iron ores in the area. Mineral related technologies were the emphasis of the research and education programs in the university and the Institute of Mineral Research was the focus of all mineral programs and was the center of mining research of Michigan state government before it was transferred to the university. In the Institute, I was able to broaden my knowledge to many other mineral technologies ranging from exploration, mining, mineral processing, metallurgy, and related environmental issues. In 1989, the Institute was renamed as the Institute of Materials Processing to bring in several materials programs such Plasma Coating, Hot and Cold Isostatic Press, Ceramics, and Powder Metallurgy and I was promoted as the Senior Research Scientist and the Research Manager to lead to the original mineral program of the Institute. In 1992, I was further promoted as the Director of the Institute of Materials Processing and oversaw all the mineral and material programs. Through this opportunity, I was able to learn many material processing technologies and expand my capacity of expertise. I was asked to join the Mining Engineering Department in 1996 for an academic appointment and served as the Chair of that department in 1997 while still serving as the Director of the Institute. Because of my research accomplishment expanded greatly into the metallurgical and materials side and the merge of the Institute with the Materials Science and Engineering Department, I was transferred to the Materials Department in 2004 and continued my Director position until 2012. The Department of Chemical Engineering and the Department of Civil and Environmental Engineering have also recruited me as their adjunct professor for my research overlap and achievements in their areas.

I am a continuous learner and have never feared to face the challenges of new research topics. I am able to combine my knowledge in mineral, mining, mineral processing, metallurgy, materials, electrical engineering, and environment to create new concepts for solving problems in the complicated world. I am able to put together a team composed of various disciplines to conduct a problem-oriented research.

I have been involved with many professional societies, especially TMS and SME. I have served as the chair of the Process Mineralogy Committee and the Secondary Materials and Environmental Committee of MPD in SME and the chair of the Materials Characterization Committee and the Pyrometallurgy Committee of EPD in TMS. I have initiated and organized or co-organized many international symposia for the two societies and edited or co-edited more than 20 books, especially the series of Materials Characterization proceedings and the International High-Temperature Metallurgical Process proceedings.

The prestigious AIME/TMS/SME James Douglas Gold Medal Award, the TMS EPD Technology Award, the TMS EPD Distinguish Service Award, the 1000 Talents Award of China, the China Non-Ferrous Society Award, and the MTU Bhakta Rath Research Award are some of the awards I received. I have also brought my graduate student team to win the Championship of the International Automotive Solutions Competition on Plastics Recycling, jointly sponsored by American Plastics Council (the major plastic companies), Vehicle Recycling Partnership (the big three), and Society of Automotive Engineers. My undergraduate senior design team is also the winner of the university-wide Champion. In addition to more than hundreds of industrial and government research projects, my research team won the US DOE Grand Challenge on Hydrogen Storage and the US DOE Grand Challenge on Steel Production of the Next Generation. My consulting activities included the Chief Energy and Environment Adviser for Wuhan Iron and Steel Corp, a Global Fortune 500 company, the Chief Technical Adviser of EST Technology Co., and the Chief Science Adviser of Futianbao Environmental Protection Technology Company. I understand what industry needs and can bridge the academic and industry well.

References

1. Hwang JY, Kullerud G, Takayasu M, FriedlaenderFJ, and Wankat PC (1982) Selective seeding for magnetic separations. IEEE Trans Mag 18:1689–1691
2. Hwang JY, Kullerud G, FriedlaenderFJ, Takayasu M (1987) Ultrafine particle processing: alunite beneficiation. AIME-SME Trans 280:1961–1964
3. Mendenhall GD, Geng Y, Hwang JY (1996) Optimization of long-term stability of magnetic fluids from magnetite and synthetic polyelectrolytes. J Colloid Int Sci 184:519–526
4. Hwang JY, Takayasu M, FriedlaenderFJ, Kullerud G (1984) Application of magnetic susceptibility gradients to magnetic separation. J ApplPhys 55:2592–2594
5. Hwang JY (2002) Adsorption of surfactant dispersed nanometer magnetite. J Minerals Mater Char Eng 1(2):132–140
6. Hwang JY (1990) Fine coal cleaning with advanced magnetic enhancement technology. In: Sixth coal preparation utilization, and environmental control conference, proceedings, pp 290–297
7. Hwang JY, Shi S, Sun X, Zhang Z, Wen C (2013) Electric charge and hydrogen storage. Int J Energy Res 37(7):741–745
8. Li X, Hwang JY, Shi S, Sun X, Zhang Z (2010) Effects of electric potential on hydrogen adsorption. Carbon 48(3):876–880
9. Xu Z, Wang H, Hwang JY (2007) Complex carbon nanotube superstructures synthesized with natural mineral catalysts. Carbon 45(4):873–879
10. Sun X, Hwang JY, Shi S (2010) Hydrogen storage in mesoporous metal oxides with catalyst and external electric field. J Phys Chem C 114(15):7178–7184
11. Li X, Hwang JY, Shi S, Sun X, Zhang Z (2010) Effect of piezoelectric material on hydrogen adsorption. Fuel Process Technol 91(9):1087–1089
12. Hwang JY, Sun X (2012) Removal of Ions from Water with Electrosorption Technology. In: Drelich J, Hwang JY, et al (eds) Water in Mineral Processing.SME Pub, pp 87–95
13. Chen Z, Huang D, Hwang JY (2019) Effect of styrene addition on chemically induced grafting of 4-vinylbenzyl chloride onto low-density polyethylene for anion exchange membrane preparation. Polym Int 68(5):972–978
14. Hwang JY, Popko D, Li B, Drelich J, Bagley S (2008) Antimicrobial property of copper stamp sand. JOM 61(2):288
15. Hwang JY, Liu X, Kesling B (1994) Sand reclamation for a steel foundry. AFS Trans 101:807–815
16. Hwang JY, Liu X, Hozeska TJ, Kesling B (1995) Reclamation of foundry sand using microwave selective heating. AFS Trans 102:789–795
17. Hwang JY, Song XM (1997) Replacing Al powder with Al slag or recycled foil in cellular concrete. JOM 49(8):29–30
18. Hwang JY, Huang X, Gillis J, Hein A, Popko D, Tieder R, McKimpson M (1999) Separation and utilization technologies of low NOx Ash. In: Proceedings: 13th international symposium on use and management of coal combustion by-products, vol 1, pp 19–1 to 19–22

19. Hwang JY, Huang X, Hein AM (1994) Synthesizing mullite from beneficiated fly ash. JOM 46(5):36–39
20. Song XM, Hwang JY (2001) Mechanical properties of wood fiber/recycled tire rubber composites. Forest Prod J 51(5):45–51
21. Hwang JY, Jeong M (2001) Separation and quantification of hazardous wastes from abrasive blast media. J AOAC Int 84(3):1–7
22. Hwang JY, Huang X (2006) New steel production technology with microwave and electric arc heating. In: Kongoli F, Reddy RG (eds) Advanced processing of metals and materials, vol 5. TMS, pp 251–261
23. Hwang JY, Huang X, Qu S, Wang Y, Shi S, Caneba G (2006) Iron oxide reduction with conventional and microwave heating under CO and H_2 atmospheres. In: Howard SM et al (eds) EPD congress 2006. TMS, pp 219–227
24. Peng Z, Hwang JY, Mouris J, Hutcheon R, Huang X (2010) Microwave penetration depth in materials with non-zero magnetic susceptibility. ISIJ Int 50(11):1590–1596
25. Sun X, Hwang JY, Huang X (2008) The microwave processing of electric arc furnace dust. JOM 60(10):26–30
26. Peng Z, Hwang JY, Andriese M, Bel W, Huang X, Wang X (2011) Numerical simulation of heat transfer during microwave heating of magnetite. ISIJ Int 51(6):884–888
27. Peng Z, Hwang JY, Andriese M (2013) Design of double-layer ceramic absorbers for microwave heating. Ceram Int 39:6721–6725
28. Peng Z, Hwang JY, Andriese M (2014) Maximum sample volume for permittivity measurements by cavity perturbation technique. IEEE Trans Instrum Measure 63(2):450–455
29. Peng Z, Hwang JY, Andriese M (2012) Absorber impedance matching in microwave heating. ApplPhys Express 5(2):077301–077301–3
30. Peng Z, Hwang JY, Andriese M (2012) Magnetic loss in microwave heating. ApplPhys Express 5(2):027304–027304–3
31. Peng Z, Hwang JY, Park C-L, Kim B-G, Onyedika G (2012) Numerical analysis of heat transfer characteristics in microwave heating of magnetic dielectrics. Metallurgical Mater Trans 43(3):1070–1078
32. Peng Z, Hwang JY (2015) Microwave-assisted metallurgy. Int Mater Rev 60(1):30–63
33. Jiang W, Hwang JY, Hao S, Zhang Y (2018) Effect of carbon coating on magnetite reduction. In: Hwang JY et al. (eds) 9th international symposium on high-temperature metallurgical processing. Springer, pp 447–454

Experimental Research on the Pellets of Brazilian Iron Concentrate

Lin Enyu and Zhao Qiang

Abstract In this paper, the method of preparing green balls from Brazilian iron concentrate was studied. The hematite and two kinds of magnetite concentrate were the main raw materials, the effects of the dosage of bentonite, pelletizing moisture, pelletizing time, particle size, and the ratio of the two magnetite concentrates on the green balls' strength were studied. When using 1.0–1.3% Indian bentonite, the pelletizing moisture of 9.5–10%, the pelletizing time of 12–15 min, the mineral powder particle size under the condition of −325 mesh 85–90%, it is suitable for pelletizing. And the falling strength of the green balls is 3.0 times 0.5 m^{-1}, the compressive strength is 16–23 N/P, the decrepitation temperature of the green balls is above 350 °C. After blending with the magnetic concentrate, the decrepitation temperature of the green balls rises, but the strength decreases, and the amount of bentonite needs to be increased.

Keywords Green ball · Hematite · Magnetite · Pelletizing · Bentonite

Introduction

China's steel output has been ranked first in the world in recent years [1]. With the rapid development of China's iron and steel technology, blast furnace charge optimization has become an important means for steel companies to reduce costs and save energy. In the process of iron and steel production, compared with the sintering process, pelletizing has the advantages of low energy consumption and less pollution. With the increasingly strict national requirements for environmental protection and the rapid development of belt roasters, the proportion of pellets entering the blast furnace has increased year by year [2–4].

As the reserves of magnetite in the world are declining and prices are rising, more and more manufacturers use hematite as a raw material [5]. Hematite has the advantages of low price, high iron content, and low impurity content, but the pellets

L. Enyu · Z. Qiang (✉)
Zhongye Changtian International Engineering Co., Ltd., Changsha 410205, China
e-mail: 13319532520@163.com

B. Li et al. (eds.), *Materials Engineering—From Ideas to Practice: An EPD Symposium in Honor of Jiann-Yang Hwang*, The Minerals, Metals & Materials Series,
https://doi.org/10.1007/978-3-030-65241-8_2

produced from hematite have lower strength and poor ball formation [6–11]. In order to improve the green ball strength, it is necessary to add an adhesive. Bentonite is currently the most widely used inorganic binder in the production of iron and steel enterprises. It has the characteristics of low cost, easy availability, sufficient resources, good spheroidization, and good burst resistance [12–14]. Besides, the addition of bentonite can also increase the strength of the green pellets.

This section studies the effects of the dosage of bentonite, pelletizing moisture, pelletizing time, mineral particle size, and the ratio of the two magnetite concentrates on green ball strength and thermal stability, and optimizes the relevant parameters for the next step of drying and preheating and roasting process provides high-quality green balls.

Experimental

Iron Concentrate

There are three types of iron concentrate used in this experiment: Brazilian MBR hematite and two domestic magnetite concentrates.

The chemical composition of MBR hematite and magnetite concentrate are shown in Table 1. The physical properties of iron concentrate are shown in Table 2. The particle size composition of MBR before and after grinding and the particle size composition of magnetite concentrate are shown in Table 3.

It can be seen from Table 1 that the FeO content of MBR hematite is only 0.19%, the iron grade is higher (66.79%), the silicon content is lower, the content of other impurities such as P, S, and non-ferrous metal elements is very low, and the particle size is fine, −0.074 mm reaches 91.14%, which is a more suitable for pelletizing.

The FeO content of the two magnet concentrates is above 24%, and w(TFe)/w(FeO) is less than 2.7, which are primary magnetite concentrates. As the main raw material of pellets, magnetite concentrate is conducive to the consolidation of pellets. When the magnetite is oxidized to hematite, it will be accompanied by structural changes, and the lattice changes in the oxidation process and the atoms on the surface of the new crystal have a higher migration ability, which is conducive to the formation of crystal bonds between adjacent particles and improves the strength of the pellets. In addition, the oxidation of magnetite to hematite is an exothermic reaction. The heat energy it emits is almost half of the total heat consumption of roasted pellets, which is helpful to make up for the lack of internal heat of roasting hematite pellets. However, when using magnetite, it must be ensured that the magnetite is fully oxidized in the pellets. If the pellets are not sufficiently oxidized in the early stage of preheating and high-temperature roasting, the pellet will have a double-layer structure or react with gangue SiO_2 at high-temperature calcination zone to produce a low melting point compound, a liquid slag phase appears inside the pellets, and concentric cracks also occur during cooling and shrinking. Both these forms will

Table 1 Chemical composition of iron concentrate/%

Mineral	TFe	FeO	SiO_2	Al_2O_3	CaO	MgO	K_2O	Na_2O	ZnO	PbO	CuO	S	P	LOI
MBR	66.79	0.19	2.47	0.82	0.10	0.038	0.071	0.014	0.014	0.017	0.0043	0.017	0.024	0.64
Magnetite 1	66.33	28.28	2.62	2.01	1.44	0.83	0.097	0.070	0.0016	0.0001	0.013	0.30	0.022	1.03
Magnetite 2	64.39	25.13	4.31	0.86	1.86	0.78	0.013	0.015	0.039	0.0097	0.0015	0.066	0.007	0.77

Table 2 Some physical properties of iron concentrate

Mineral	Bulk density/g cm^{-3}	True density/g cm^{-3}	Porosity/%	Static accumulation angle/°	Static sphericity index	Specific surface area/cm^2 g^{-1}
MBR	2.1719	4.55	52.27	34.34	0.19	520
Magnetite 1	2.2460	4.57	50.85	35.71	0.242	1650
Magnetite 2	1.862	4.71	60.47	33.28	0.651	1875

Table 3 Iron concentrate particle size composition/%

Granularity/mm	+0.150	−0.150 + 0.074	−0.074 + 0.043	−0.043
MBR	1.63	7.23	29.55	61.59
Magnetite 1	0.30	8.00	26.05	65.65
Magnetite 2	1.85	4.30	32.30	61.55

affect the strength of the pellets and even deteriorate the reducing properties of the pellets.

Bentonite

The bentonite required for the pelletizing test is Indian bentonite. The main chemical components are shown in Table 4, and the physical properties are shown in Table 5. It can be seen from Table 4 that Indian bentonite has high iron content and low silicon content, which is beneficial to iron extraction and silicon reduction. Indian bentonite is sodium-based bentonite with a Na_2O content of up to 2.92%. From the point of view of its physical properties, Indian bentonite has the highest amount of blue absorption, and the content of montmorillonite exceeds 100%, which is related to the adsorption of methylene blue reagent by other clay minerals in bentonite and the increase of blue absorption. In addition, the swelling capacity and water absorption index are

Table 4 Analysis of chemical composition of bentonite/%

Variety	TFe	SiO_2	Al_2O_3	CaO	MgO	K_2O	Na_2O	S	P	LOI
Indian	11.89	48.27	15.86	1.62	2.69	0.091	2.92	0.059	0.046	10.32

Table 5 Physical properties of bentonite

Variety	Blue adsorption/g $(100\ g)^{-1}$	Colloid index/%	Swelling capacity/mL g^{-1}	Water absorption (2 h)/%	−0.074 mm/%	−0.043 mm/%	pH
Indian	44.5	99.5	37.0	552.12	98.41	65.32	10.05

also very high. The swelling capacity is 37 mL/g, the 2-h water absorption rate is 552.12%, and the colloid index is 99.5%. It is a high-quality bentonite. Bentonite is a commonly used binder for iron ore pellets, which can improve the green ball strength and burst temperature, but its silicate composition will reduce the iron grade of the pellet ore [15]. Therefore, it is generally desirable to reduce the amount of bentonite added on the premise of satisfying green ball strength and thermal stability.

Results and Discussion

The raw materials used for making pellets must go through artificial ingredients first, and then mix them before sending them to pelletizing.

Each batch of the small-scale experiment is based on 5 kg of iron-containing raw materials, and each batch of the expanded test is based on 20 kg of iron-containing raw materials, and bentonite is added at a certain ratio. When performing manual mixing, adding a certain amount of pre-wetting moisture to make the moisture of the mixture lower than the proper moisture of the green ball.

The pelletizing experiments are carried out in a disc pelletizer. Its main technical parameters are diameter $\Phi = 1200$ mm, revolving speed 38 r/min, side height h = 150 mm, and inclination angle $\alpha = 47°$. In the small-scale test, the amount of each batch is 5 kg when making balls, and 20 kg for the expanded test. The green balls made are manually selected, and the 8–16 mm green balls are regarded as qualified green balls. Some green balls are sampled to determine their compressive strength, drop strength, decrepitation temperature and green ball moisture, and the remaining green balls are dried for subsequent preheating and roasting experiments.

Pelletizing Experiments

In this section, the effects of bentonite ratio, fineness of iron ore fines, mixing ratio, mixture moisture, and pelletizing time on green ball strength and thermal stability were studied, and related parameters were optimized for the next drying, preheating, and roasting process to provide high-quality green balls.

Effect of Bentonite Dosage

The fixed MBR particle size is 85% at −325 mesh, the pelletizing moisture is 10% ± 0.2%, and the pelletizing time is 12 min. The effect of Indian bentonite on the pelletizing experiments of the entire MBR mine was investigated (Table 6).

Table 6 The influence of Indian bentonite on the quality of green balls at 85% of MBR −325 mesh

Bentonite dosage/%	Drop strength/times $0.5\ m^{-1}$	Compressive strength/N P^{-1}	Decrepitation temperature/°C
0	1.3	8.78	>600
0.7	2.4	21.83	538
1.0	3.2	23.70	358
1.3	3.7	24.40	342

It can be seen that for the pelletizing experiment of the whole MBR mine, under the above-fixed conditions, at a particle size of 85% at −325 mesh, without adding bentonite pellets, the green ball strength is only 1.3 times $0.5\ m^{-1}$.

From the experiments of adding bentonite, Indian bentonite can ensure the falling strength of green balls up to 3 times $0.5\ m^{-1}$ at a lower amount of bentonite, and the compressive strength of green balls is above 15 N/P, with the amount of bentonite added increased, the decrepitation temperature of the green ball decreased.

Considering that the addition of 1% of bentonite will reduce the iron grade of the finished pellets by about 0.5–0.6%, therefore, on the premise of meeting the quality requirements of green balls, the amount of bentonite should not be added too much. In the subsequent pelletizing experiments, Indian bentonite is the main component, and the dosage is 1.0–1.3%.

Effect of the Moisture of Pelletizing

With a fixed pelletizing time of 12 min, the particle size of MBR is 85% at −325 mesh, and the amount of Indian bentonite is 1.0%. The effect of pelletizing moisture on the quality of the whole MBR ore pellet is investigated. The results are shown in Table 7.

It can be seen from Table 7 that as the moisture of the pelletizing increases, the drop strength and compressive strength of the raw pellets first increase and then decrease, but in the moisture range of 9.5–10.5%, the difference in green balls strength is not obvious. The decrepitation temperature decreases with increasing moisture. This is because as the moisture of the green balls increases, the capillary gravitational energy increases, and the plasticity of the material increases, especially Indian bentonite

Table 7 Effect of moisture on green ball quality of MBR 85% at −325 mesh

Green balls moisture/%	Drop strength/times $0.5\ m^{-1}$	Compressive strength/N P^{-1}	Decrepitation temperature/°C
9.48	3.0	20.53	524
10.04	3.2	23.70	358
10.55	2.6	22.24	335

Table 8 Effect of pelletizing time on green ball quality of MBR 85% at −325 mesh

Pelletizing time/min	Drop strength/times $0.5\ m^{-1}$	Compressive strength/N P^{-1}	Decrepitation temperature/°C
9	2.6	21.45	510
12	3.2	23.70	358
15	3.0	24.24	324
18	2.5	26.58	296

has excellent performance, so the drop strength and compressive strength of the green ball increase, and the decrepitation temperature will decrease instead. Within the range of 9.5–10.0% of the pelletizing moisture tested, the drop strength of the green balls of MBR mine is 3.0 times $0.5\ m^{-1}$, the compressive strength is 20–23 N/P, and the decrepitation temperature of the green ball is 358–584 °C, the green ball quality indexes can basically meet the production requirements. Therefore, the suitable pelletizing moisture is in the range of 9.5–10.0%.

Effect of Pelletizing Time

With a fixed particle size of MBR is 85% at −325 mesh, the moisture of pelletizing is 10% ± 0.2% and the amount of Indian bentonite is 1.0%. The effect of pelletizing time on the quality of the whole MBR ore pellet is investigated. The results are shown in Table 8.

It can be seen from the table that under the given conditions when the pelleting times are 12 min and 15 min, the drop strengths of the green ball reaches 3.2 times/0.5 m and 3.0 times/0.5 m, respectively, and the compressive strength of the green ball also reaches 23.70–24.24 N/P, the decrepitation temperatures are 358 °C and 324 °C, respectively. With the extension of the pelletizing time, the drop strength and decrepitation temperature of the green balls decrease, and the compressive strength increases. Therefore, the pelletizing time should not be longer than 15 min, and the appropriate pelletizing time is 12–15 min.

Effect of Grinding Fineness

The fixed pelletizing moisture is 10% ± 0.2%, and the pelletizing time is 12 min. The effect of different grinding sizes of MBR on the quality of green balls is investigated. The experiment results are shown in Table 9.

It can be seen from Table 9 that with the increase in grinding fineness, the content of −0.043 mm particle size increases, the drop strength of the green balls increases significantly, and the decrepitation temperature shows a downward trend. Looking

Table 9 Effect of grinding fineness on green ball quality of MBR 85% at −325 mesh

Grinding fitness −325 mesh/%	Drop strength/times 0.5 m^{-1}	Compressive strength/N P^{-1}	Decrepitation temperature/°C
80	2.1	18.15	568
85	3.2	23.70	358
90	4.1	21.07	384
95	5.0	19.94	360

at the indicators of green balls strength and decrepitation temperature, it is more suitable for the MBR ore powder to have a particle size of 85–90% at −325 mesh after fine grinding. If the particle size is too fine, the grinding cost will increase too much. In addition, the effect of the appropriate fine particle size composition on the green ball strength may exceed the effect of the fine powder particle size on the green ball strength. Therefore, the follow-up experiments mainly use 90% fine powder of −325 mesh fineness.

Blending and Pelletizing Experiments

Fix 90% of MBR at −325 mesh as hematite raw material, pelletizing time 12 min, pelletizing moisture 10 ± 0.2%, investigate the blending pelletizing experiments of MBR ore and magnetite concentrate. During the test, the effects of different types and amounts of bentonite on the performance of the green balls of the magnet concentrate were also investigated. At the same time, the effect of the proportion of the magnet concentrate on the green balls is also investigated.

Ore Blending Experiment with Magnetic Concentrate 1

Table 10 shows the results of the pelletizing test when using Indian bentonite. From the results above, under the same amount of bentonite, as the proportion of magnetic concentrate 1 increases, the drop strength of the green ball decreases significantly. On the other hand, the decrepitation temperature of the green ball is greatly increased compared with the full hematite pelleting. Comprehensive consideration, magnetic concentrate 1 is not suitable as a raw material for MBR ore mixing. If magnetic concentrate 1 is added to the mixture, the amount of bentonite needs to be appropriately increased to maintain the drop strength of the green ball.

Table 10 Effects of different ratios of magnetic concentrate 1 on the quality of 90% green balls of MBR at −325 mesh

Ratio of magnetic concentrate 1/%	Bentonite dosage/%	Drop strength/times 0.5 m^{-1}	Compressive strength/N P^{-1}	Decrepitation temperature/°C
0	1.0	4.1	21.07	384
10	1.0	3.0	16.45	422
20	1.0	2.0	15.08	512
30	1.0	1.9	17.64	499
20	1.3	3.1	15.79	438

Ore Blending Experiment with Magnetic Concentrate 2

The results of ore-making and pelletizing test with magnetic concentrate 2 are shown in Table 11. When using Indian bentonite, it can be clearly found that with the increase of the ratio of magnetic concentrate 2, the drop strength of the green ball drops greatly, in the case of 1.0% Indian bentonite dosage, the drop strength of the green ball with the magnetic concentrate 2 is less than 3.0 times 0.5 m^{-1}, only by increasing the amount of bentonite to maintain the drop strength of the green ball.

According to the experiment of adding magnetic concentrate 1 and magnetic concentrate 2, when using Indian bentonite, as the proportion of magnetic concentrate increases, the drop strength of green balls decreases, which shows that bentonite has a certain adaptability to iron ore. The difference in surface charge also affects the performance of the green ball.

Table 11 Effects of different ratios of magnetic concentrate 2 on the quality of 90% green balls of MBR at −325 mesh

Ratio of magnetic concentrate 1/%	Bentonite dosage/%	Drop strength/times 0.5 m^{-1}	Compressive strength/N P^{-1}	Decrepitation temperature/°C
0	1.0	4.1	21.07	384
10	1.0	2.9	22.16	379
20	1.0	2.6	18.43	489
30	1.0	2.2	20.07	426
20	1.3	3.4	19.55	435

Conclusion

Different bentonites have different effects on the quality of green balls. From the perspective of improving the drop and compressive strength of green balls, Indian > Fanchang > Husi; and from the perspective of the impact on the decrepitation temperature of green balls, Husi > Fanchang > Indian. Under laboratory conditions, the dosage of domestic bentonite is 1.5–2.0%, and the dosage of Indian bentonite is 1.0–1.3%. Comprehensively considering the drop strength and decrepitation temperature of green balls, the suitable pelletizing moisture is in the range of 9.5–10.0%, and the suitable pelletizing time is 12–15 min. At this time, the drop strength of the green balls is 3.0 times 0.5 m^{-1}. The compressive strength is 20–23 N/P. The decrepitation temperature of the green ball is above 350 °C. Looking at the indexes of green ball strength and bursting temperature, it is more suitable for the MBR ore powder to have a particle size of 85–90% at −325 mesh after fine grinding. When the magnetic concentrate is added, the drop strength of the green ball with the magnetic concentrate 2 is better than that of the green ball with the magnetic concentrate 1. Compared with the full hematite for the magnetic concentrate, when using Indian bentonite, it is necessary to increase the amount of bentonite to meet the requirement of the drop strength of the green ball.

References

1. Huang ZC, Zhang YB, Zhu SP (2004) Study on mainly made from hematite based oxidized pellets with addition of some magnetite. Iron Steel 59(4):9–13
2. Wang HF, Pei YD, Zhang CX (2016) Green development of sintering/pellet procedure in China iron and steel industry. Iron Steel 1:1–7
3. Xu MX, Zhang YL (2017) Analysis of pellet technology and production of China in 21st century. Sinter Pelletizing 37(2):25–30
4. Liu H, Xiang XP, Gan MY (2019) Experimental study on preparation of pellets using high MgO-bearing iron ore concentrate. Sinter Pelletizing 2:33–38
5. Zhu DQ, Shen WJ, Pan J (2011) Research on influence of the alkalinity and the fuel flux to the intensity of hematite pellets. Metal Mine 6:104–108
6. Yin ZG, Tang HQ, Ma XY (2017) Study on intensifying the preparation of oxidized pellets from Iranian hematite concentrates. 2:38–42
7. Fan XH, Yuan XL (2006) Effect of particle size distribution of concentrate on palletizing quality. Chin J Nonferrous Metals 11:1965–1970
8. Zheng HX, Wang Q, Pan XF (2003) Mechanisms of oxidation of magnetite pellet. Sinter Pelletizing 5:13–16
9. Chen YM, Zhang YB (2005) Study on crystallization rule of oxidized pellet. Res Iron Steel 3:10–12
10. Chen YM, Li J (2007) Crystal rule of Fe_2O_3 in oxidized pellet. J Central South Univ 1:70–73
11. Wang CA, Xiong SA, Zhu DQ (2002) Effect of high pressure roller mill pre-treating concentrates on green ball properties. Sinter Pelletizing 6:12–15
12. Li JL, Zhu JW, Gong ZY (2020) Economic evaluation on technologies for composite binders for iron ore pellets. Angang Technol 2:18–24
13. Hou EJ, Weng XY, Fan XH (2019) Study on strengthening preparation of oxide pellets with fibration composite bentonite. Sinter Pelletizing 4:40–44

14. Wu X, Peng XM, Chen YH (2016) New high efficient adhesive iron concentrate oxidized pellets experiment. Modern Mining 9:78–81
15. Li JL, Liu WS, Ren W (2012) Experiment study on decreasing bentonite consumption of pellets by adding high iron red mud. Sinter Pelletizing 3:31–34

Recent Progress in Microwave-Assisted Pyrometallurgy at Central South University

Liancheng Wang, Zhiwei Peng, Jie Wang, Wenxing Shang, Qiang Zhong, Mingjun Rao, Guanghui Li, and Tao Jiang

Abstract Microwave-assisted pyrometallurgy has many distinguished advantages over traditional pyrometallurgical processes. In the past 5 years, the researchers at Central South University (CSU) has been focusing on the use of microwave energy for intensifying various pyrometallurgical processes for higher processing efficiency, enhanced metal recovery, lower energy consumption, and better environmental benefits. This study reviewed the efforts of the researchers at CSU in microwave-assisted pyrometallurgy for treatment of wastes, with an aim to offer a useful guide for improving the technology.

Keywords Microwave · Pyrometallurgy · Recycling · Secondary resources · Extraction

Introduction

Microwaves are electromagnetic waves with a frequency range from 300 MHz to 300 GHz, or equivalently with wavelengths of 1 mm–1 m. Among them, 915 and 2450 MHz are the most widely used frequencies in the industry and household for heating purpose [1]. Because microwaves can deliver electromagnetic energy directly into the interior of samples and convert it into heat, they are featured by the volumetric thermal effect which contributes to a higher heating efficiency. Moreover, as microwave energy conversion depends on the dielectric and magnetic properties of the raw materials, they have selective thermal effects, which can further improve processing efficiency [2]. Due to its economic and environmental advantages in association with volumetric heating and selective heating, microwave has various

L. Wang · Z. Peng (✉) · J. Wang · W. Shang · Q. Zhong · M. Rao · G. Li · T. Jiang
School of Minerals Processing and Bioengineering, Central South University, Changsha 410083, Hunan, China
e-mail: zwpeng@csu.edu.cn

National Engineering Laboratory for High Efficiency Recovery of Refractory Nonferrous Metals, Changsha 410083, Hunan, China

B. Li et al. (eds.), *Materials Engineering—From Ideas to Practice: An EPD Symposium in Honor of Jiann-Yang Hwang*, The Minerals, Metals & Materials Series,
https://doi.org/10.1007/978-3-030-65241-8_3

applications in metallurgy [3]. It has been widely used for intensifying drying, calcination, grinding, leaching, reduction, roasting, smelting, sintering, etc. In particular, because of its low cost and high environmental benefits [4], it plays a prominent role in the treatment of metallurgical wastes (e.g., metallurgical dust and slag) and other secondary resources [5, 6].

In recent years, there has been obvious progress in the research of microwave-assisted pyrometallurgy at the Central South University (CSU). The purpose of this paper was to review the relevant efforts of CSU on the treatment of various wastes in the past 5 years.

Microwave-Assisted Treatment of Wastes

Metallurgical Dusts

The output of steel plant dusts is about 10% of the output of crude steel [7]. Among various metallurgical dusts, electric arc furnace (EAF) dust and blast furnace (BF) dust are the two representative ones. Generally, the EAF dust contains 30–45 wt% iron, 3–40 wt% zinc, and a large amount of gangue minerals and heavy metal elements, including lead and chromium [7–11]. The dust has been classified as hazardous waste, which should be recycled or treated properly. According to the chemical composition, it has three types, namely low-zinc dust (<4 wt% Zn), medium-zinc dust (4–20 wt% Zn), and high-zinc dust (>20 wt% Zn). Unlike EAF dust, BF dust is a kind of valuable dust produced in ironmaking [10]. BF dust usually contains 25–50 wt% iron, 1–5 wt% zinc, and 15–25 wt% carbon [10], and can be used as an excellent carbon resource. The elemental compositions of EAF and BF dusts used at CSU are shown in Table 1 [8–10].

Because microwave can selectively heat the spinel phases in the EAF dust (especially franklinite $ZnFe_2O_4$) and biochar is a good reducing agent, derived from renewable waste biomass, with strong microwave absorption capacity and less hazardous impurities, combining microwave and biochar was expected to greatly improve reduction performance of EAF dust. The CSU research group evaluated the effects of different reducing agents, including biochar, on the reduction of EAF dust based on both microwave and conventional heating [9]. The results showed that under the same reducing conditions, microwave-assisted self-reduction of composite briquettes of EAF dust and biochar was superior to conventional reduction. After microwave treatment at 1050 °C for 15 min, the iron metallization degree reached 99%, which was much higher than the conventional process (56.0%). Compared with coal and coke, biochar presented a much higher efficiency, in association with its well-developed pore structure which promoted microwave absorption and carbon gasification. The results confirmed that microwave intensified self-reduction of composite briquettes of EAF dust and biochar had the advantages of high efficiency, low energy consumption, and good environmental friendliness. Table 2 compares the results [9].

Table 1 Elemental compositions of EAF and BF dusts used at CSU

Element	EAF dust			BF dust (wt%)
	High-zinc EAF dust (wt%)	Medium-zinc EAF dust (wt%)	Low-zinc EAF dust (wt%)	
Zn	30.50	5.65	1.63	0.015
Fe	27.61	52.54	22.46	42.47
Ca	2.11	2.39	21.03	6.29
Na	1.05	2.02	1.19	0.08
Pb	0.37	0.12	0.059	0.008
Si	1.35	1.10	3.86	2.15
K	1.02	1.12	0.31	0.22
Mn	1.85	1.77	3.50	0.13
S	0.37	0.56	0.71	0.16
Cl	2.60	0.54	0.18	0.01
Mg	1.23	0.76	2.61	1.22
Al	0.43	–	5.60	1.66
Cr	0.12	0.56	0.23	0.006

Table 2 Comparison between reduction processes using different heating methods and reducing agents

Parameters	Iron metallization degree (wt%)		Reduction conditions
	Conventional	Microwave	
Biochar	56.0	>99.0	1050 °C, 15 min
Coal	31.0	97.0	1050 °C, 15 min
Coke	41.2	73.4	1050 °C, 15 min

Preparing direct reduced iron (DRI) using EAF dust and biochar as raw materials under microwave irradiation has also been studied [8]. Based on the analysis of microwave penetration depth and reflection loss of the raw materials, the propagation and attenuation behaviors of microwave in the mixture of biochar and EAF dust were determined for producing qualified composite briquettes of EAF dust and biochar to achieve rapid microwave reduction with superior DRI product. Under the optimal conditions of the mass ratio of biochar to EAF dust of 0.25, reduction temperature of 1050 °C, dwell time of 15 min, and heating rate of 25 °C /min, the iron metallization degree of reduction product reached 94.7%, with volatilization of 99.6% zinc and 92.9% lead. After magnetic separation, DRI with total iron content of 91.32 wt% and recovery of 87.36% was obtained and could be used as a burden for EAF steelmaking. The low-temperature microwave reduction reduced the dissolution of carbon in the γ-Fe and inhibited the formation of Fe_3C. It was found that the "lens effect" of microwave promoted the reduction with enhanced aggregation of metallic particles.

For clarifying the microwave heating mechanism of microwave-assisted reduction of EAF dust using biochar, the CSU research group examined the microwave response of composite briquette of EAF dust–biochar in different stages of reduction [6]. The changes in microwave absorption capacity of the briquette were determined by measuring the variations of permittivity and permeability during the reduction process. According to the temperature–magnetic moment curve, the magnetic loss considerably affected the heating process. As the temperature rose, the magnetic moment decreased. The first derivative of temperature–magnetic moment curve showed that the Curie point was around 504 °C. When the temperature was below 504 °C, the dielectric polarization and magnetic loss controlled the temperature rise. With further increase of temperature, the magnetic loss tended to disappear.

Based on the variation of dielectric parameters with temperature, the heating process was divided into three stages. In stage I, the dielectric polarization and magnetic loss controlled the reduction of the composite briquette. When the temperature increased to 504 °C, the spontaneous magnetization of the composite system dropped to unity. The dielectric constant and dielectric loss factor remained relatively stable. In stage II, up to 800 °C, the dielectric constant increased due to the reduction of magnetite and zinc ferrite. The volatilization of zinc and the newly generate metallic iron mainly affected the change of the dielectric loss factor. At this stage, the dielectric polarization contributed greatly to heating. In stage III, due to the massive production of metallic iron and the volatilization of zinc, the composite briquette was filled with electronic charges, which promoted the microwave response. Above 800 °C, electronic conduction (e.g., conduction loss) was the main factor of heating.

Except for the reducing agent, the composite system structure affected the microwave-assisted reduction of EAF dust. A core–shell composite pellet of EAF dust and biochar (as reducing agent) was built with the C/Fe mass ratio of 0.24 of pellet, the C/Fe mass ratio of 0.30 of pellet core, and diameter of pellet core of 6–8 mm [10]. After reduction at 1050 °C for 15 min, the iron metallization degree of core–shell pellet was higher than homogeneous pellet (97.97% versus 94.96%), with volatilization of 99.52% zinc and 98.49% lead. The unique core–shell structure realized the selective distribution of carbon, which not only contributed to the efficient carbon gasification but also improved the heat diffusion in the pellet, thereby accelerating various chemical reactions. This result was consistent with the findings obtained in microwave-assisted reduction of core–shell iron ore–biochar composite pellet with the selective distribution of biochar [12]. Further work on using BF dust as reducing agent is ongoing.

Metallurgical Slags

Laterite ore is widely used as a raw material in ferronickel production in China, and the annual output of ferronickel slag has exceeded 30 million tons [13–15]. Although the slag can be used to produce cement aggregates, inorganic polymers and geopolymers, its utilization percentage is still around 10% [13]. The main components of

Table 3 Main chemical composition of ferronickel slag

Component	FeO	SiO_2	MgO	CaO	Al_2O_3	NiO	Cr_2O_3
Content (wt%)	7.39	48.29	30.95	2.40	4.04	0.06	2.11

ferronickel slag are FeO, MgO, Al_2O_3, and SiO_2, which exist in the forms of enstatite, diopside, olivine, forsterite, and non-crystalline minerals [13–15]. Table 3 shows the chemical composition of a typical ferronickel slag [13].

The CSU research group proposed to prepare refractory materials from ferronickel slag by microwave sintering in the presence of magnesia, to achieve the value-added utilization of the slag. Under the combined effects of microwave and magnesia, the main phase in the slag was transformed from original olivine to forsterite due to the evolution of enstatite and magnesium iron chromate spinel. The presence of magnesia would modify the phases and microstructures of refractory material during microwave sintering. Under the optimal conditions of sintered magnesia addition of 20 wt%, sintering temperature of 1350 °C, and sintering time of 20 min, the resulting material had superior quality with the refractoriness of 1730 °C, bulk density of 2.80 g/m^3, apparent porosity of 1.6%, and compressive strength of 206.62 MPa. This method not only reduced the sintering time by six times compared with conventional sintering but also increased the refractoriness by 70 °C [13].

For further improving the properties of refractory material, the CSU research group prepared refractory materials by microwave sintering of ferronickel slag with additions of both magnesia and alumina. The results showed that the presence of alumina improved forsterite grain refinement and uniformity. In addition, the formations of high melting point phases, including magnesia aluminum spinel ($MgAl_2O_4$), magnesium iron aluminate spinel ($MgFe_{0.6}Al_{1.4}O_4$), and magnesium aluminum chromate spinel ($MgAl_{0.5}Cr_{1.5}O_4$), were promoted. After microwave sintering of mixture of ferronickel slag, 20 wt% sintered magnesia, and 4 wt% alumina at 1250 °C for 20 min, a high-quality refractory material with refractoriness of 1790 °C was obtained. This method provided a more efficient and environmental-friendly approach for the value-added treatment of ferronickel slag [14].

Apart from using ferronickel slag to prepare refractory materials, recovering heavy metals, such as chromium (2.11 wt% Cr_2O_3 in the ferronickel slag), from the ferronickel slag is also of significance. The CSU research group recovered chromium by microwave roasting of the mixture of ferronickel slag and Na_2O_2 with the mass ratio of 1, followed by water leaching. The main chemical reactions are given as follows:

$$Mg_2SiO_4 + Fe_2SiO_4 + 3Na_2O_2 = 2Na_2MgSiO_4 + Na_2Fe_2O_4 + O_2 \quad (1)$$

$$2Cr_2O_3 + 2Na_2O_2 = 4NaCrO_2 + O_2 \quad (2)$$

After microwave roasting at 600 °C for 20 min, the Cr_2O_3 in the ferronickel slag was converted into $NaCrO_2$, which was highly soluble in water. Meanwhile, Mg_2SiO_4

and Fe_2SiO_4 were converted to Na_2MgSiO_4 and $Na_2Fe_2O_4$, which remained insoluble in water. After microwave roasting followed by water leaching, 94.21% of chromium was leached with 64.43% of sodium and 6.13% of silicon. The chromium content was reduced to 0.06 wt%. This new process contributed to a higher chromium leaching percentage compared with the conventional route (81.29%), verifying its higher processing efficiency in recovering chromium from the slag [15].

Spent Catalysts

Platinum group metals (PGMs) include platinum (Pt), palladium (Pd), rhodium (Rh), ruthenium (Ru), iridium (Ir), and osmium (Os). They have been widely used in electronic devices, catalytic converters, thermocouples, etc. Due to the low abundance of PGMs in the earth's crust, recovering PGMs from secondary resources, e.g., spent catalysts from automobile catalytic converters, becomes an important subject [16].

The pyrometallurgy can effectively separate precious metals with high selectivity. For improving recovery of PGMs from spent catalysts, more efforts should be made to find suitable collectors and flux, and appropriate intensification processes should be used to recover the metals [17]. The CSU research group investigated control of the composition of slag from smelting of spent catalysts by adjusting the viscosity (additive CaO) and melting temperature using the thermodynamic software FactSage 7.1 (Thermfact/CRCT and GTT-Technologies, Montreal, Canada and Herzogenrath, Germany) [18]. The thermodynamic results showed that the addition of Na_2CO_3 could accelerate the formation of $NaAlO_2$ and $Na_2Si_2O_5$ in the slag and decrease the melting temperature of spent catalysts. Moreover, $Na_2B_4O_7$ increased the flowability of the slag. The presence of $Na_2B_4O_7$ would cause the formation of $NaAlSiO_4$, $NaAlSi_3O_8$, and $(Al_2O_3)_9(B_2O_3)_2$ below 1000 °C, which indicated the similar function of $Na_2B_4O_7$ to Na_2CO_3. By the synergistic effect of Na_2CO_3 and $Na_2B_4O_7$, the release of PGMs from the oxide carriers in the spent catalysts could be achieved, further decreasing the usage of various collectors. In order to verify the results of thermodynamic calculations, microwave smelting has been carried out. When the spent catalysts were heated to 1100 °C for 1.5 h, the collection percentages of Pt, Pd, and Rh in nickel buttons exceeded 90%. The addition of alkali metal salts promoted the migration and separation of PGMs from spent catalysts. Replacing CaO with Na_2CO_3 and $Na_2B_4O_7$ was able to reduce the viscosity and the smelting temperature (reduced by more than 200 °C) of spent catalysts. All results verified that microwave smelting with the addition of sodium salt was an effective method for recovering PGMs from spent catalysts.

Conclusions

This article reviewed the representative progress of CSU in the field of microwave-assisted pyrometallurgy in the past 5 years. Microwave was applied to the treatment of various wastes, including EAF/BF dust, ferronickel slag, and spent catalysts. Because of the good coupling performance with EAF dust and biochar, microwave intensified the self-reduction of EAF dust-biochar composite briquettes, achieving excellent reduction in a short time. In addition, changing the structure of composite system could further improve the reduction performance. As for the treatment of ferronickel slag, microwave sintering was adopted for producing high-quality refractory materials and for recovering chromium from the slag in the presence of Na_2O_2, minimizing the slag threat to the environment. For spent catalysts, microwave was found useful and efficient in its treatment. All the efforts of the research group at CSU demonstrated that the microwave process is a low-cost, low energy consumption, and environmentally friendly metallurgical method for waste treatment.

Acknowledgements This work was partially supported by the National Natural Science Foundation of China under Grants 51774337, the Science and Technology Planning Project of Hunan Province, China under Grant 2019RS2008, the Natural Science Foundation of Hunan Province, China, under Grant 2017JJ3383, the Hunan Provincial Innovation Foundation for Postgraduate under Grant CX20190230, and the Fundamental Research Funds for the Central Universities of Central South University under Grants 2019zzts174 and GS201910533550.

References

1. Chen G, Li L, Tao C, Liu Z, Chen N, Peng J (2016) Effects of microwave heating on microstructures and structure properties of the manganese ore. J Alloy Compd 657:515–518
2. Samouhos M, Taxiarchou M, Hutcheon R, Devlin E (2012) Microwave reduction of a nickeliferous laterite ore. Miner Eng 34:19–29
3. Huang J, Xu G, Liang Y, Hu G, Chang P (2020) Improving coal permeability using microwave heating technology: a review. Fuel 266:117022
4. Ye Q, Peng Z, Li G, Lee J, Liu Y, Liu M, Wang L, Rao M, Zhang Y, Jiang T (2019) Microwave-assisted reduction of electric arc furnace dust with biochar: an examination of transition of heating mechanism. ACS Sustain Chem Eng 7:9515–9524
5. Bobicki ER, Pickles CA, Forster J, Marzoughi O, Hutcheon R (2020) High temperature permittivity measurements of selected industrially relevant ores: review and analysis. Miner Eng 145(1):106055
6. Kingman SW, Rowson NA (1998) Microwave treatment of minerals-a review. Miner. Eng. 11(11):1081–1087
7. Peng Z, Hwang JY (2015) Microwave-assisted metallurgy. Int Mater Rev 60(1):30–63
8. Ye Q, Li G, Peng Z, Augustine R, Pérez MD, Liu Y, Liu M, Rao M, Zhang Y, Jiang T (2020) Microwave-assisted self-reduction of EAF dust-biochar composite briquettes for production of direct reduced iron. Powder Technol 362:781–789
9. Ye Q, Li G, Peng Z, Yang L, Zheng L, Wang J, Shang W, Anzulevich A, Rao M, Li G, Jiang T (2019) Microwave-assisted self-reduction of composite briquettes of zinc ferrite and carbonaceous materials. Powder Technol 342:224–232

10. Wang L, Peng Z, Ye Q, Wang L, Augustine R, Lee J, Liu Y, Liu M, Rao M, Li G, Jiang T (2020) Preparation of metallized pellets from blast furnace dust and electric arc furnace dust based on microwave impedance matching. In: 11th international symposium on high-temperature metallurgical processing, pp 569–579
11. Lin X, Peng Z, Yan J, Li Z, Hwang JY, Zhang Y, Li G, Jiang T (2017) Pyrometallurgical recycling of electric arc furnace dust. J Clean Prod 149:1079–1100
12. Ye L, Peng Z, Wang L, Anzulevich A, Bychkov I, Tang H, Rao M, Zhang Y, Li G, Jiang T (2018) Preparation of core-shell iron ore-biochar composite pellets for microwave reduction. Powder Technol 338:365–375
13. Tang H, Peng Z, Gu F, Ye L, Hwang JY, Rao M, Li G, Jiang T (2020) Alumina-enhanced valorization of ferronickel slag into refractory materials under microwave irradiation. Ceram Int 46(5):6828–6837
14. Peng Z, Tang H, Augustine R, Lee J, Tian W, Chen Y, Gu F, Zhang Y, Li G, Jiang T (2019) From ferronickel slag to value-added refractory materials: a microwave sintering strategy. Resour Conserv Recy 149:521–531
15. Peng Z, Wang L, Gu F, Tang H, Rao M, Zhang Y, Li G, Jiang T (2020) Recovery of chromium from ferronickel slag: a comparison of microwave roasting and conventional roasting strategies. Powder Technol 372:578–584
16. Demarco JO, Cadore JS, Veit HM, Madalosso HB, Tanabe EH, Bertuol DA (2020) Leaching of platinum group metals from spent automotive catalysts using organic acids. Miner Eng 159:106634
17. Peng Z, Li Z, Lin X, Tang H, Ye L, Ma T, Rao M, Zhang Y, Li G, Jiang T (2017) Pyrometallurgical recovery of platinum group metals from spent catalysts. JOM 69:1553–1562
18. Peng Z, Li Z, Lin Z, Ma Y, Zhang Y, Zhang Y, Li G, Jiang T (2018) Thermodynamic analysis of smelting of spent catalysts for recovery of platinum group metals. In: 9th international symposium on high-temperature metallurgical processing, pp 215–223

Production and High-Ratio Application of Iron Ore Pellets in Shougang

Gele Qing, Minge Zhao, Gang An, Kai Wang, Xiaojiang Wu, and Zhixing Zhao

Abstract In order to reduce the fuel rate of blast furnace and promote green development, Shougang Jingtang Steel company has built two straight grate indurating machines with a productivity of 8 million tons of iron ore pellets. In this paper, the production technologies of self-fluxed pellets with hydrated lime were studied and the high-quality pellets were successfully produced on the indurating machine. Hydrated lime has a good binder property and decreases the amount of bentonite to obtain low silica pellet. The basicity is 1.15 and SiO_2 content is 2.0% in the pellet. The low-silica self-fluxed pellets were used in three large blast furnaces with 5500 m^3 in Shougang Jingtang. The proportion of pellets in burden increased from 28 to 55%, slag rate of blast furnace decreased from 280 to 215 kg/tHM, and the fuel rate decreased from 500 to 480 kg/tHM.

Keywords Pellet · Basicity · Blast furnace · Slag rate · Fuel rate

Introduction

In the long process of the iron and steel industry, the blast furnace ironmaking and converter steelmaking processes play a dominant role [1]. The main charges for blast furnace ironmaking are pellets and sinter. Due to iron ore resources, domestic blast furnace ironmaking is mainly based on sinter, and the average ratio of sinter in blast furnace is above 80% [2]. However, in terms of the process, the energy consumption of the pelletizing process is about 45% of that in the sintering process, and the emission of flue gas, SO_2, NOx, and particulate matter is about 55% of that in the sintering

G. Qing (✉) · Z. Zhao
Shougang Research Institute of Technology, Beijing 100043, China
e-mail: qinggele_68@163.com

M. Zhao
Shougang Group, Beijing 100041, China

G. An · K. Wang · X. Wu
Shougang Jingtang United Iron & Steel CO., Ltd., Hebei 063200, China

B. Li et al. (eds.), *Materials Engineering—From Ideas to Practice: An EPD Symposium in Honor of Jiann-Yang Hwang*, The Minerals, Metals & Materials Series,
https://doi.org/10.1007/978-3-030-65241-8_4

process. At the same time, the SiO_2 content of pellets can be reduced to below 3.0%, or even to the level of 2.0% comparing to the level of above 4% in the sinter. If the blast furnace uses a high proportion of low-silica pellets, the amount of blast furnace slag and fuel consumption can be greatly reduced so that CO_2 emissions can be reduced eventually. However, it is difficult to reduce the SiO_2 content of sintered ore to below 4.0%, so the amounts of slag and fuel consumption in a blast furnace using a high proportion of sinter are difficult to be reduced to a relatively low level [3]. Therefore, increasing the proportion of pellets used in blast furnace ironmaking is of great significance for reducing pollutants and CO_2 emissions in the ironmaking process. According to the demand of slag basicity, it is necessary to increase the basicity and reduce the SiO_2 content of pellets to increase the ratio of pellets in blast furnace.

To promote low-carbon and green development, Shougang Jingtang Company has built pellet production lines in the second phase of the project. After construction, the ironmaking equipment includes three big blast furnaces with a volume of 5500 m^3, two sintering machines with an area of 500 m^2, and three straight grate indurating machines with an area of 504 m^2. In order to achieve high-proportion pellet utilization to reduce slag and fuel consumption of blast furnace, this paper studies the appropriate flux for the production of low-silica and high-basicity pellets, the roasting process parameters and metallurgical measures for basic pellets. Meanwhile, the production and application in Shougang Jingtang Steel company have been described.

Raw Materials and Experimental Methods

Physical and Chemical Properties of Raw Materials Used in the Test

The iron ore concentrate used in the experiments is a high-grade, low-silica magnetite concentrate with the iron grade 69.65%, SiO_2 content 1.69%, and the FeO content 29.35%. In order to analyze the suitable flux for producing low-silica basic pellets, two kinds of fluxes as limestone and slaked lime were tested and studied. The binder for the pellets was ordinary bentonite. Table 1 shows the chemical compositions of the test materials. The CaO content and the burning loss of slaked lime are 70.26% and 22.26%, respectively, while the CaO content and the burning loss of limestone powder are 52.47% and 43.48%, respectively.

Research Methods

The batching and mixing were carried out according to the proportioning plan. Green pellet was prepared in pelletizing disc with a diameter of 800 mm and pelletizing time

Table 1 Chemical compositions of test materials (mass%)

Item	TFe	SiO_2	CaO	P	S	FeO	Al_2O_3	MgO	K_2O	Na_2O	Burning loss
Iron ore	69.65	1.69	0.35	0.004	0.17	29.35	0.33	0.48	0.043	0.087	−2.66
Limestone	–	1.31	52.47	0.009	0.037	–	0.12	1.22	0.12	0.088	43.48
Slaked lime	–	1.19	70.26	0.008	0.031	–	0.34	3.76	0.034	0.008	22.26
Bentonite	–	63.35	2.49	0.022	0.010	–	14.2	2.45	1.23	1.72	9.31

of 10 min. After pelletizing the green pellets with the size of 10–16 mm were screened out for roasting test. The roasting test was carried out in the experimental pot. The compressive strength, reduction swelling, and reducibility were tested according to ISO standards. FactSage software was used to estimate the liquid phase formation temperature and amount of liquid with different SiO_2 contents and basicities.

Results and Discussion

The Influence of Different Fluxes on the Strength of Green Balls

Pelletizing experiments were carried out by adding 1%, 2%, 3%, 4%, and 5% limestone powder or slaked lime to concentrate powder, and the ratio of bentonite was 1.5% and 0.5%, respectively. Figure 1 shows the change of green ball drop strength with the ratio of limestone powder and slaked lime when bentonite content was 1.5%. With the increase of the slaked lime proportion, the drop strength of the green balls

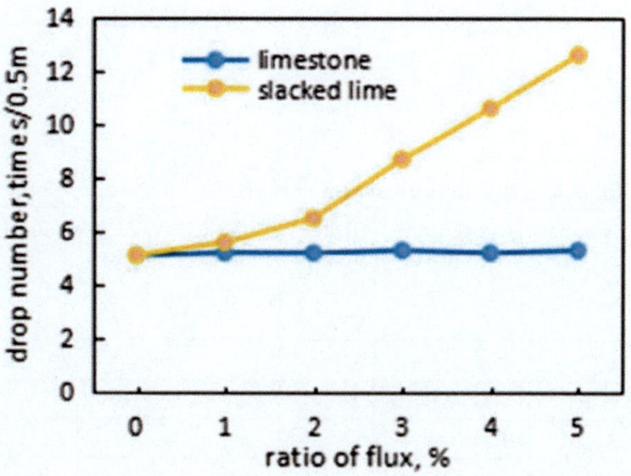

Fig. 1 Drop strength of green ball (bentonite: 1.5%). (Color figure online)

increases significantly. When the proportion of slaked lime is 2.0%, the average drop strength of the green balls reaches 6.6 times/0.5 m, which can meet the production requirements. However, the drop strength of pellets with limestone powder does not change much. Under the ratio of 1.5% bentonite, the average number of drops of the pellets is almost fixed around 5.0 times/0.5 m. Figure 2 shows the green ball drop strength when the ratio of bentonite is reduced to 0.5%. It can be seen from Fig. 2 that when the ratio of bentonite is reduced to 0.5%, the drop strength of pellets with slaked lime is still relatively high. When the proportion of slaked lime is 2.0%, the average number of drops of green balls reaches 5.5 times/0.5 m, which can satisfy production requirements. However, it could not meet the production requirements when the drop strength of green balls with limestone powder dropped to about 3.3 times/0.5 m. It shows that the cohesiveness of slaked lime is beneficial to improve the ball-forming performance and green ball index of mineral powder.

When slaked lime was used, the ratio of bentonite can be reduced, and this is beneficial to reduce the SiO_2 content of the pellets. Meanwhile, due to the high CaO content of slaked lime, the ratio of slaked lime is lower than that of limestone when the same basicity is reached. Table 2 shows the ratio of flux and bentonite and the composition of pellets. When adding 2.7% slaked lime and 0.5% bentonite, the SiO_2

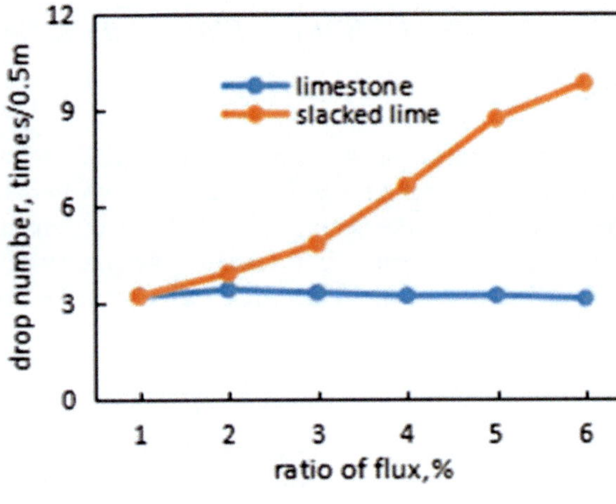

Fig. 2 Drop strength of green ball (bentonite: 0.5%). (Color figure online)

Table 2 Ratio of flux and bentonite and the compositions of pellets

Number	Concentrate ratio (%)	Slaked lime ratio (%)	Limestone ratio (%)	Bentonite ratio (%)	Compositions of pellets		
					SiO_2 (%)	CaO (%)	Basicity
1	100	3.8		1.5	2.59	3.06	1.20
2	100	2.7		0.5	1.98	2.30	1.20
3	100		5.3	1.5	2.60	3.11	1.20

content of pellets is 1.98% and the basicity is 1.2. While using limestone powder, 5.3% limestone powder needs to be added to achieve the basicity of 1.2 and the SiO_2 content stays around 2.60%. Therefore, the use of slaked lime to produce basic pellets can increase the basicity and reduce the SiO_2 content.

Study on Roasting Temperature of Low-Silica Fluxed Pellets

FactSage software was used to calculate the liquid phase formation temperature of pellets under different basicities when the SiO_2 content was 3.0, 2.5, and 2.0%, as shown in Fig. 3. When the SiO_2 content of pellets is 3.0%, and the basicity is 0.3–0.8, the liquid phase formation temperature is relatively low (1168–1171 °C). When the basicity is 0.8, the liquid phase formation temperature increases to about 1243 °C. When the basicity is 1.0, the liquid phase formation temperature reached 1262 °C. When the SiO_2 content of pellets is 2.0% and the basicity is 0.4–0.6, the liquid phase formation temperature is 1168–1175 °C. When the basicity is above 0.7, the liquid phase formation temperature is above 1240 °C, while the liquid phase formation temperature is 1248 °C when the basicity is 1.0. This shows that when producing low-silica high basicity pellets, a certain amount of liquid phase can be formed by increasing the roasting temperature.

Figure 4 shows the change in liquid phase formation amount of pellets when roasting at 1250 °C and different SiO_2 contents with basicity. In Fig. 4, the liquid phase formation of pellets with high SiO_2 content is relatively high. For example, when the basicity is 0.5, the liquid phase formation of pellets with SiO_2 content of 3.0% is 6.09%, while the amount of liquid phase formed is 2.37% for pellets with

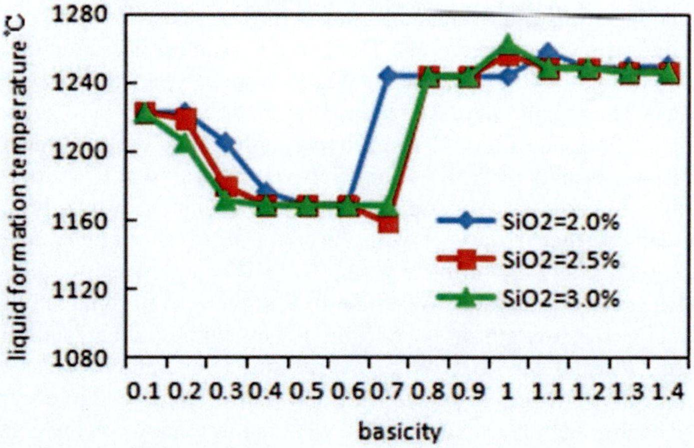

Fig. 3 Liquid phase formation temperature of pellets with different basicities and SiO_2 contents. (Color figure online)

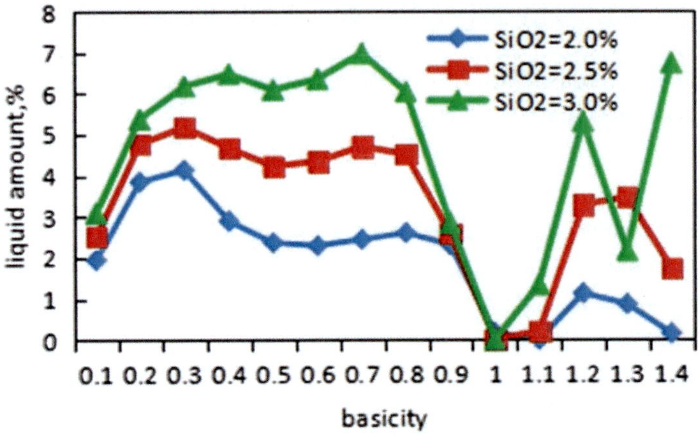

Fig. 4 Liquid phase formation amount of pellets with different basicity and SiO_2 contents. (Color figure online)

SiO_2 content of 2.0%. When the basicity is 1.2, the liquid phase formation amount of pellets with 3.0% SiO_2 content and 2.0% SiO_2 content are 5.3% and 1.12%, respectively. Therefore, the low-silica pellets are mainly consolidated through the oxidation and recrystallization of magnetite. If a certain amount of liquid phase is to be formed, the roasting temperature needs to be increased.

Study on the Roasting Temperature of Pellets Added with Flux

The pellets with adding 0, 1, 2, and 3.0% slaked lime were used to carry out roasting experiments at different temperatures. The pellet roasting process includes drying, preheating, roasting, soaking, and cooling. Figure 5 shows the pellet roasting temperature and time. The maximum roasting temperature of curve 1, curve 2, and curve 3 are 1240 °C, 1250 °C, and 1260 °C, respectively, at the same roasting time of 52 min. The compressive strength of the roasted pellets was measured after the roasting test. Figure 6 shows the compressive strength of pellets with slaked lime. From the test results, we can see that in curve 1 with the highest roasting temperature at 1240 °C, the compressive strength of the roasting ball shows a downward trend with the increase of the slaked lime ratio. When the roasting temperature increased to 1250 °C, the compressive strength of slaked lime pellets with adding 2 and 3% slaked lime reached 2967 and 2864 N/P. When the roasting temperature increased to 1260 °C, the compressive strength of pellets with slaked lime increased more significantly and reached an average of more than 3300 N/P. It shows that when the concentrated powder is mixed with 2–3% slaked lime, the pellet roasting temperature is best to be controlled at 1250–1260 °C, and the compressive strength of pellets can reach more than 2800 N/P.

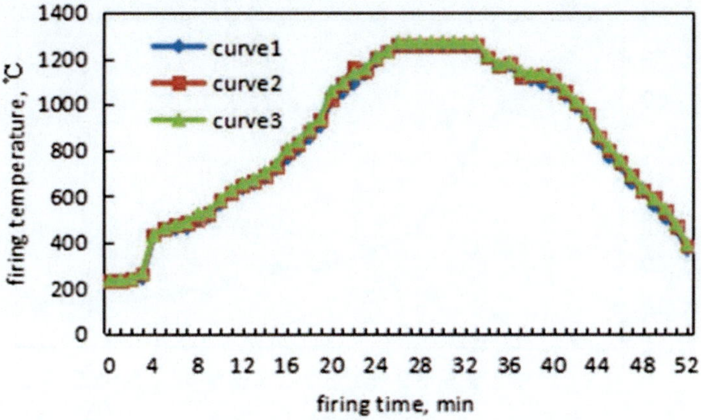

Fig. 5 Roasting temperature and time of the test. (Color figure online)

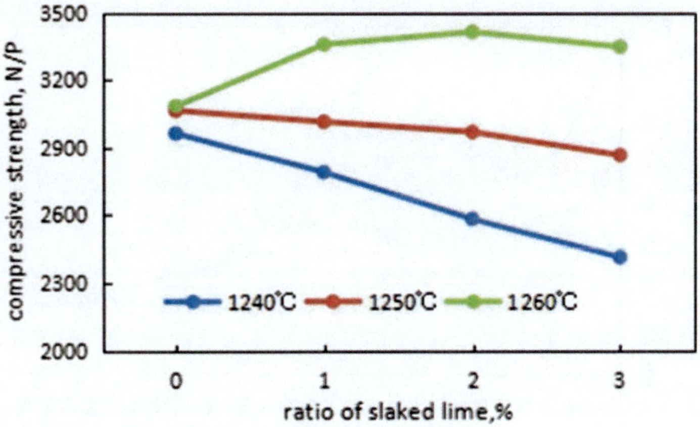

Fig. 6 Compressive strength of pellets. (Color figure online)

Research on the Reduction Swelling of Low-Silica Basic Pellets

Figure 7 shows the change of reduction swelling index of pellets with various SiO_2 contents. It can be seen that the reduction swelling index of pellets decreases with increasing SiO_2 content. The reduction swelling index of the pellets is below 20% when the SiO_2 content is higher than 3.3%, while the reduction swelling index of the pellets exceeds 20% when the SiO_2 content is less than 3.3%. When the SiO_2 content is 2.0%, the reduction swelling index is 68.5%, which cannot meet the application requirements of the blast furnace. Therefore, the reduction swelling must be controlled to produce low-silica pellets. Figure 8 shows the results of reduction swelling index of pellets roasted with different proportions of slaked lime. It can be seen from Fig. 10 that as the basicity increases, the reduction swelling index of

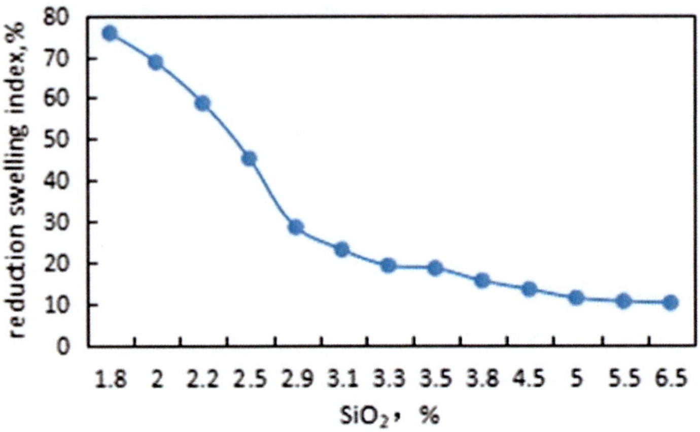

Fig. 7 Reduction swelling index changes with SiO_2 content. (Color figure online)

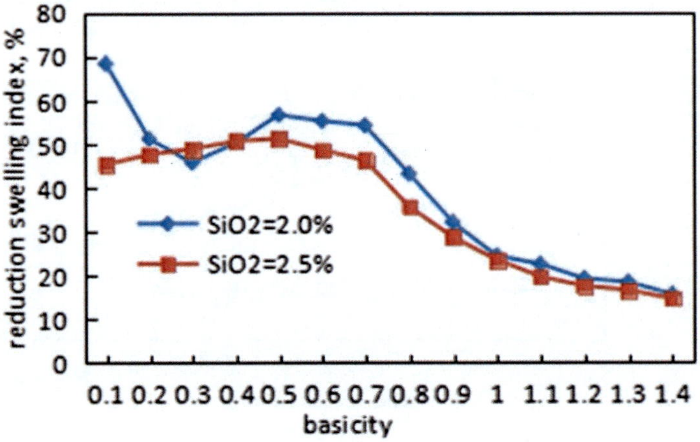

Fig. 8 Reduction swelling index changes with basicity. (Color figure online)

low-silica pellets decreases. When the SiO_2 content is 2.5%, the basicity increases to 1.1 and the reduction swelling index can be reduced to 19.7%. When the SiO_2 content is 2.0%, the basicity increases to 1.2 and the reduction swelling index is reduced to 19.3%.

From the perspective of the microstructure of the pellets after roasting, the primary phase in the low-silica pellets is hematite when the basicity is 0.8, accompanied with less liquid phase, as shown in Fig. 9. Therefore, iron whiskers are prone to appear during reduction, resulting in abnormal swelling of pellets, as shown in Fig. 10. After the basicity increased to 1.2, a certain amount of liquid phase appears inside the pellets, as shown in Fig. 11. The liquid phase is beneficial to inhibit the development

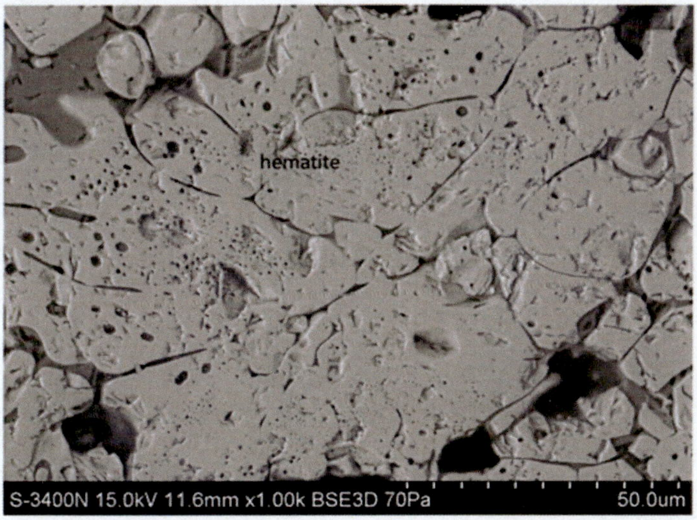

Fig. 9 SEM image of roasted pellet (basicity: 0.8)

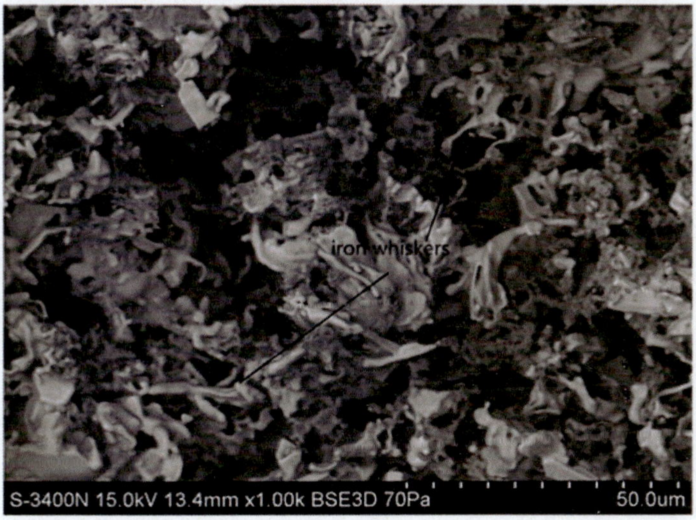

Fig. 10 SEM images of reduced pellet (basicity: 0.8)

of iron whiskers, and thus the reduction swelling index of pellets is controlled, as shown in Fig. 12. It shows that the low silica pellets can increase the basicity to control the reduction swelling.

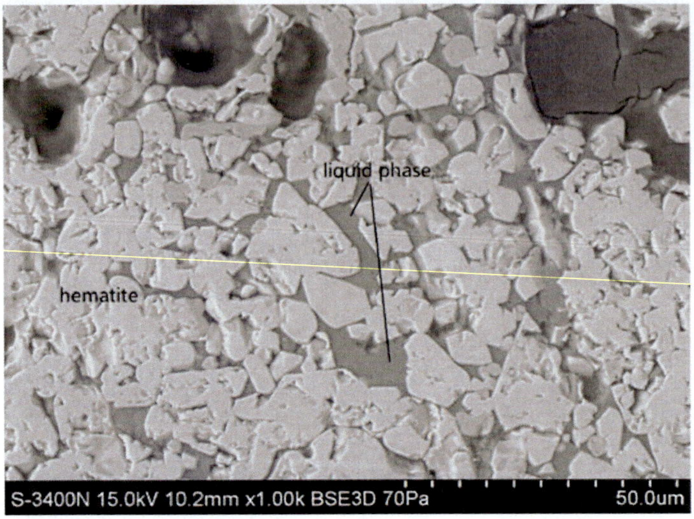

Fig. 11 SEM image of roasted pellet (basicity: 1.2)

Fig. 12 SEM images of reduced pellet (basicity: 1.2)

Production of Low-Silica Basic Pellets

In order to produce low-silica basic pellets, Shougang Jingtang Steel Company has established two slaked lime production lines with a capacity of 250,000 tons. Table 3 shows the indicators of pellets produced by Shougang Jingtang Company. The SiO_2

Table 3 The main indicators of pellets

Pellet composition (%)			Compressive strength (N/P)	Reduction swelling index (%)	Reduction degree (%)
TFe	SiO_2	B2			
66.0	2.0	1.18	3256	16.5	86.3

Table 4 Blast furnace burden structure and indicators

	Burden structure (%)			Blast furnace indicators	
	Sinter	Pellet	Lump ore	Slag rate (kg/t)	Fuel rate (kg/t)
Base period	67	28	5	280	500
Usage period	40	55	5	215	480

content of the pellet is 2.0%, basicity is 1.1–1.2, compressive strength is 3256 N/P, the reduction swelling index is 16.5%, and the reduction degree is 86.3%.

Application of High Proportion of Pellets in Blast Furnace

After the completion of the second stage projects of the Shougang Jingtang Company, the output of pellets increased significantly to 12 million tons annually. And the pellet ratio of burden in three big blast furnaces increased from 28 to 55% and the percentage of sinter decreased from 67 to 40%. Table 4 shows the changes in blast furnace charge structure. After the proportion of pellets increased, the amount of blast furnace slag reduced from 280 to 215 kg/t, and the fuel ratio is reduced from 500 kg/t to about 480 kg/t.

Conclusions

(1) The slaked lime has good binding properties, which is beneficial to improve the spherization performance and green ball index of the slaked lime and reduce the pellet bentonite ratio and SiO_2 content. Moreover, the slaked lime has a high CaO content, low heat absorption of decomposition, and high pellet compressive strength. This will be beneficial to the production of low-silica basic pellets.

(2) When the SiO_2 content of pellets is low and the basicity is high, the liquid phase formation temperature is relatively high. When the SiO_2 content is 2.0% and the basicity of the pellets is above 1.1, a certain amount of liquid phase can be formed under the temperature of 1250–1260 °C. It will help to control the reduction swelling of low-silica pellets.

(3) Shougang Jingtang Company uses slaked lime to produce basic pellets. The pellets have high-grade and low SiO_2 content, which is beneficial to reduce the amount of blast furnace slag and fuel consumption. The proportion of pellets in the three 5500 m^3 blast furnaces has increased from 28 to 55%, the blast furnace slag volume has been reduced from 280 to 215 kg/t, and the fuel ratio has been reduced from 500 kg/t to about 480 kg/t.

References

1. Wang W (2018) Increase the proportion of pellets in blast furnace burden and promote energy saving and emission. In: National annual meeting of blast furnace ironmaking, vol 10, no 18, pp 16–20
2. Wu G, Bian M, Shen F (2006) Application of basic magnesium bearing pellets and study on reasonable charge structure. Iron Steel 41(12):19–22
3. Qing G, Wu K, Liu H, Yuan X, Tian Y (2015) Effect of roasting temperature on reduction swelling of low silica magnesium bearing pellet. Chin J Nonferrous Metals 25(10):2905–2912

Part II
Polymer Materials and Processes

Blending of Polystyrene-Block-Poly(Ethylene-Ran-Butylene)-Block-Polystyrene with Polyethylene-Graft-Polystyrene for Cation Exchange Membrane Preparation with Enhanced Properties

Zhichao Chen, Jiann-Yang Hwang, Yong Shi, Di Huang, and Weigang Zhao

Abstract A typical cation exchange membrane can be prepared from graft polymerization of styrene onto low-density polyethylene followed by sulfonation reaction. It was found that the thermoplastic elastomer polystyrene-block-poly(ethylene-ran-butylene)-block-polystyrene (PS-PEB-PS) can be blended with polyethylene-graft-polystyrene (PE-g-PS) in toluene. A novel composite cation exchange membrane was prepared by sulfonating the membrane cast from the mixture solution of PE-g-PS and PS-PEB-PS. Compared with the non-blended cation exchange membrane, the blending of 10% PS-PEB-PS enhanced the ion exchange capacity, water uptake, and Mullen burst strength of the cation exchange membrane by 18.3%, 9.5%, and 26.3%, respectively. The blending also lowered the thermal expansion rate by 73.4%, making the membrane more thermally stable over temperature increase.

Keyword Cation exchange · Thermoplastic elastomer · Composite membrane

Introduction

Electrodialysis (ED) is an advanced separation technique that uses ion exchange membrane to separate ions under an electrical field [1]. Ion exchange membrane is widely used in desalination of seawater and treatment of industrial wastewater that contains metal cations [2, 3]. Proton exchange membrane is known to be a core part of fuel cell that transports protons from anode to cathode [4, 5]. Ion exchange membranes are mostly functionalized polymers that selectively allow ions to pass through [6]. They can be categorized into two types based on the charge they carry.

Z. Chen (✉) · Y. Shi · W. Zhao
Shaanxi Futianbao Environmental Protection Technology, Xi'an, China
e-mail: zhichaochen1989@sina.com

J.-Y. Hwang · D. Huang
Michigan Technological University, Houghton, MI, USA
e-mail: jhwang@mtu.edu

B. Li et al. (eds.), *Materials Engineering—From Ideas to Practice: An EPD Symposium in Honor of Jiann-Yang Hwang*, The Minerals, Metals & Materials Series,
https://doi.org/10.1007/978-3-030-65241-8_5

Cation exchange membranes carry negative charge on their structure. The negative charges are mostly from functional groups such as SO_3^-, COO^-, PO_3H^-, and $C_6H_4O^-$ that allow transportation of cations through the membranes [7]. In contrast, anion exchange membranes have positive charge that comes from functional groups such as $N(Me)_3^+$, NH_3^+, and many others, and allow permeation of anions [5, 8].

Styrene and polystyrene are widely used for the synthesis of ion exchange resins and membranes due to their wide availability and inexpensive prices [9, 10]. Polystyrene can be physically blended with or chemically bonded to inert polymers so that the obtained precursor membranes can maintain mechanical strength [11]. Typically, the polystyrene containing membranes are sulfonated or aminated to introduce sulfonic acid groups or trimethylammonium groups into the structure to make the cation or anion exchange membranes [12, 13].

Polystyrene-block-poly(ethylene-ran-butylene)-block-polystyrene (PS-PEB-PS) is a widely used thermoplastic elastomer which is comprised of three blocks. PS-PEB-PS has a high concentration of styrene in the polymer backbone that can be functionalized to either cation or anion exchange groups [14, 15]. As a rubber material, PS-PEB-PS has high ductility and impact resistance which makes it a potential additive material for polystyrene based ion exchange membranes. In this study, PE-graft-PS (PE-g-PS) was synthesized by grafting polystyrene onto low-density polyethylene and blended with PS-PEB-PS in toluene. A novel cation exchange membrane was prepared by sulfonating the composite membrane with chlorosulfonic acid.

Experimental

Materials

Low-density polyethylene (LDPE, Mw = 5000) beads, styrene (St, 99%), benzyl peroxide (BPO, 99%), tetrahydrofuran (THF, 99.5%), and 1,2-dichloroethane (99%) were purchased from Damao chemical reagent factory. Divinylbenzene (DVB, 80%), xylene (99%), ethanol (99.7%), and chlorosulfonic acid (99%) were supplied by Macklin. PS-PEB-PS (Mw = 118,000) was purchased from Sigma-Aldrich. All solvents were used as received.

Characterization and Evaluation of Cation Exchange Membrane

The structure, morphology, and composition of polymer membranes were analyzed by Fourier transform infrared (FTIR), scanning electron microscope (SEM), and elemental analysis. FTIR spectra were obtained by a Nicolet IS50 with universal attenuated total reflection accessories. A Zeiss Gemini SEM500 was used to observe morphology and phase separation on the surface of the membrane. The elemental composition of the membranes wes analyzed by a PerkinElmer 2400 CHNS elemental analyzer with helium as the carrier gas. The Mullen burst strength of the membrane sample was determined by a Mullen burst tester supplied by Hangzhou Zhibang automation technology.

Synthesis of PE-g-PS Copolymer

Following a known procedure, 19.2 g LDPE beads were placed in a two-neck round bottom flask with toluene and xylene [9]. The solution was purged nitrogen and heated to 90 °C for LDPE to dissolve. To initiate the copolymerization reaction, 19.2 g styrene and 0.95 g divinylbenzene were added followed by benzyl peroxide. The reaction was kept at 90 °C for 6 h for the reaction to complete. The viscous solution was poured onto a piece of glass that was heated to 120 °C to evaporate the solvents. A thin membrane was obtained and washed with THF to remove impurities.

Preparation of PE-g-PS/PS-PEB-PS Composite Membrane

The obtained PE-g-PS membrane was cut into pieces and placed in a flask with toluene, then 2.88 g PS-PEB-PS was added to the flask. The mixture was heated to 90 °C and kept for 2 h for the mixture to form a homogeneous and viscous solution. A PE-g-PS/PS-PEB-PS composite membrane was obtained in the same way described for PE-g-PS membrane.

Sulfonation of PE-g-PS/PS-PEB-PS Composite Membrane to Cation Exchange Membrane

Following methods described in the literature, sulfonation of PE-g-PS membrane or PE-g-PS/PS-PEB-PS composite membrane was carried out by soaking the membrane in 1,2-dichloroethane with 10% chlorosulfonic acid in an ice bath for 2 h [15]. After reaction, the membrane was repeatedly washed with THF and ethanol to remove solvent residue. The resulting membrane was kept in 1 M HCl at room temperature for 24 h to form H^+ form of cation exchange membrane. The membrane was then dried in an overset at 50 °C.

Determination of Ion Exchange Capacity (IEC)

The IEC of ion exchange membrane was determined by a typical titration method [16, 17]. The membranes were cut into small pieces and stirred with 1.0 M sulfuric acid solution overnight to assure that the ion exchange groups are in H^+ form. Then, the membranes were washed with distilled water to remove the excess sulfuric acid. The resulting membranes were dried in the oven at 60 °C and weighted. The membranes were then stirred with NaCl (0.5 M) solution overnight to convert to sodium form. The sodium cation will replace H^+ on the sulfonic acid group and result in acidic solution. The concentration of H^+ was determined by titration method using diluted KOH solution (0.01 M). The IEC values were calculated using the Eq. (1):

$$\text{IEC (mmol/g)} = (V_{KOH} \times C_{KOH})/W, \quad (1)$$

where V_{KOH} is the volume of KOH used in the titration, W is the weight of the dried membrane in g, and C_{KOH} is the concentration of KOH used in the experiment for the titration.

Determination of Water Content

The water content is determined by Eq. (2):

$$\text{Water content (\%)} = (W_2 - W_1)/W_1 \times 100\%, \quad (2)$$

where W_1 is the mass of the dried membrane samples and W_2 is the mass of the membrane samples after soaking in 0.5 M NaCl for 24 h and removal of the surface moisture.

Determination of Thermal Expansion Rate

The thermal expansion rate is determined by Eq. (3):

$$\text{Thermal expansion rate (\%)} = (L_2 - L_1)/L_1 \times 100\%, \quad (3)$$

where L_1 is the average size of a square sample of the dried membrane and L_2 is the average size of the square sample after soaking in water for 4 h at an elevated temperature.

Results and Discussion

FTIR and SEM Analysis

In the FTIR spectra, the original LDPE showed strong bands at 2919, 2851, 1471, and 717 cm^{-1}, corresponding to C–H bond stretching and bending in LDPE (Fig. 1). After graft polymerization of polystyrene, additional peaks were observed at 3025, 1604, and 1495 cm^{-1}. These are from C–H bond on the benzene ring of styrene [18]. The FTIR pattern does not change after blending 10% PS-PEB-PS into PE-g-PS. This is due to the fact that PS-PEB-PS are composed of polystyrene and poly(ethylene-co-butylene) which overlap with PE-g-PS in FTIR pattern. After sulfonation with chlorosulfonic acid, a new band was observed at 1125 cm^{-1} for PE-g-PS. This is from S = O stretching vibration on SO_3^- [12]. Peaks at 1003 and 1033 cm^{-1} were believed to result from bending vibration of benzene ring plane when it is affected by SO_3^-. The peak at 3395 cm^{-1} is due to bending vibration of O–H on sulfonic acid group [12]. The sulfonated PE-g-PS/PS-PEB-PS has a similar pattern in FTIR except for that peaks at 2917 and 2848 cm^{-1} are relatively stronger. This is due to the additional alkyl groups from PS-PEB-PS. The FTIR confirmed the successful graft copolymerization and sulfonation of the membranes.

In the SEM image, spheres ranging from nanometers to micrometers were observed (Fig. 2). This is consistent with Mizutani's research and our previous study on anion exchange membranes [6, 18]. This kind of microheterogeneity or phase

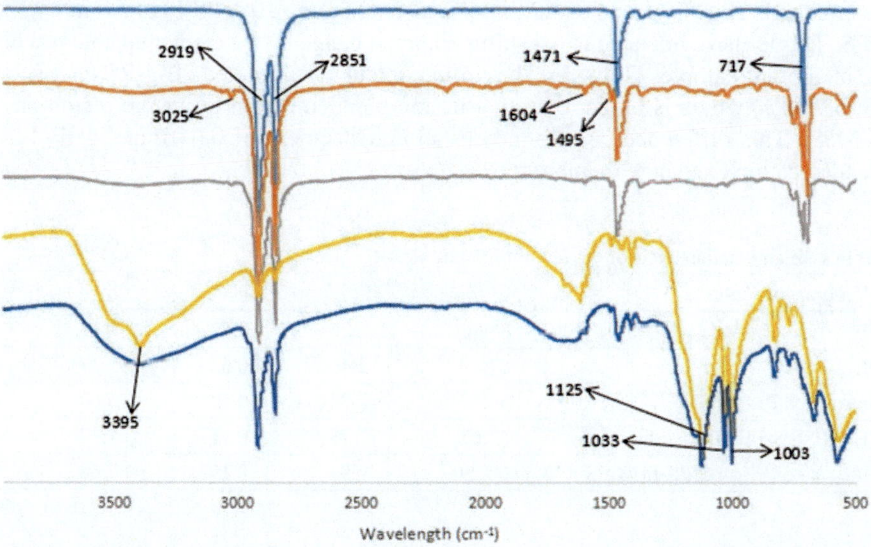

Fig. 1 FTIR spectra of membrane samples (from top to bottom: LDPE, PE-g-PS, PE-g-PS/PS-PEB-PS, sulfonated PE-g-PS, sulfonated PE-g-PS/PS-PEB-PS). (Color figure online)

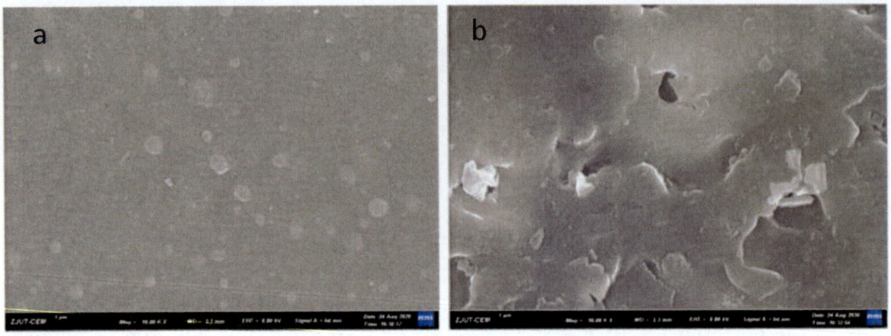

Fig. 2 SEM of PE-g-PS (**a**) and PE-g-PS/PS-PEB-PS membrane (**b**). (Color figure online)

separation is due to the incompatibility between polyethylene and polystyrene. With the addition of PS-PEB-PS, the surface of the membrane became more uneven with a gum-like structure. Microspheres still can be observed, but less than those on PE-g-PS. The sulfonated membranes did not show much difference in the morphology from copolymers before sulfonation.

IEC and Elemental Analysis

The sulfonated PE-g-PS membrane has an IEC of 1.85 mmol/g determined by titration method. This increased to 2.2 mmol/g after blending 10% PS-PEB-PS into PE-g-PS. This is also evidenced in the sulfur content change by the elemental analysis of the membrane samples (Table 1). The sulfur content in sulfonated PE-g-PS/PS-PEB-PS is 7.68%, which is higher than that in non-blended cation exchange membrane (5.64%). This difference corresponds to an enhancement of 0.64 mmol/g IEC in blended cation exchange membrane.

Table 1 Elemental analysis of the membrane samples

Membrane samples	C (%)	H (%)	N (%)	S (%)
PE	85.85	15.26	0.58	1.13
PE-g-PS	87.79	12.46	0.63	1.37
Sulfonated PE-g-PS	69.67	10.83	0.52	5.64
PE-g-PS/PS-PEB-PS	87.62	12.73	0.17	1.17
Sulfonated PE-g-PS/PS-PEB-PS	66.29	9.91	0.19	7.68

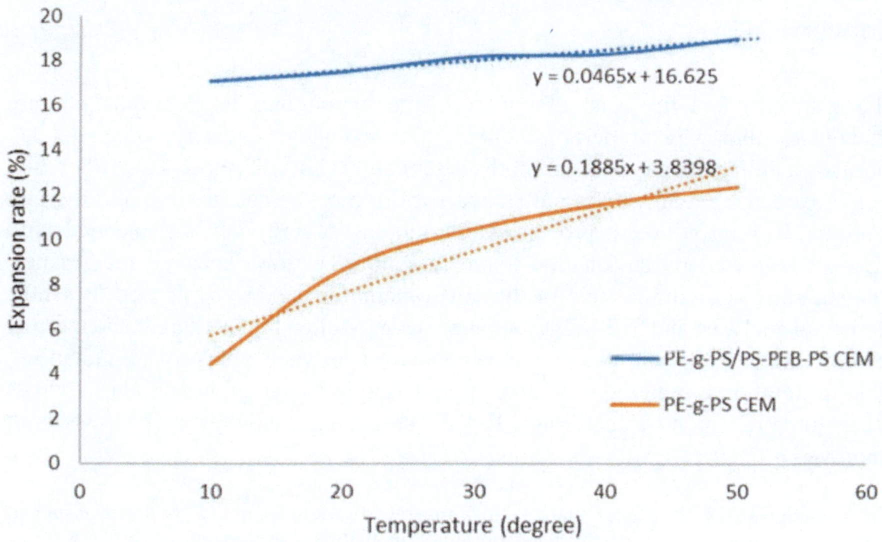

Fig. 3 Thermal expansion of PE-g-PS/PS-PEB-PS CEM. (Color figure online)

Water Content and Thermal Expansion Rate

The water content in PE-g-PS/PS-PEB-PS composite cation exchange membrane is determined to be 35.2%, which is higher than that in non-blended sample (25.7%). This is due to the higher IEC value and resultant higher hydrophilicity of the composite membrane.

From Fig. 3, we can see that the PE-g-PS/PS-PEB-PS composite cation exchange membrane expands by 17.1% upon absorbing water. This is higher than non-blended sample (4.4%). However, this blended membrane is very stable over temperature change with a slope of 0.047%/degree. This is much lower than that for the non-blended sample (0.19%) and indicates better thermal stability over temperature change for blended sample.

Mullen Burst Strength

Both membranes were measured to be 40–50 μm in thickness. The Mullen burst strength was determined to be 2388 kPa/mm for blended cation exchange membrane. This is higher than the non-blended sample (1890 kPa/mm) by 26.3%. The Mullen burst strength can probably be tuned by varying the ratio of the blended components.

Conclusion

The efficiency and life span of ion exchange membranes in electrodialysis are affected by their key properties including ion exchange capacity, water uptake, thermal stability, mechanical strength, and many others. Typical PE-g-PS based cation exchange membrane is not satisfactory in mechanical strength and thermal stability. To improve these properties, thermoplastic elastomer was blended with PE-g-PS followed by sulfonation to prepare novel cation exchange membranes. The structure and composition of the novel membranes were confirmed by FTIR, elemental analysis, and SEM. They showed better Mullen burst strength and thermal stability. Meanwhile, the ion exchange capacity and water uptake were enhanced. This straightforward method can also apply to anion exchange membranes and has great potential for manufacturing PE-g-PS-based ion exchange membranes with improved properties.

Acknowledgements We thank Futianbao Environmental Protection Co. Ltd. for financial support of this work and Michigan Technological University for analytical instruments.

References

1. Xu T (2005) Ion exchange membranes: state of their development and perspective. J Membr Sci 263:1–29
2. Chatterjee U, Jewrajka SK (2014) Amphiphilic poly(acrylonitrile)-co-poly(2-dimethylamino)ethyl methacrylate conetwork-based anion exchange membrane for water desalination. J Mater Chem A 2:8396–8406
3. Wang Q, Yang P, Cong W (2011) Cation-exchange membrane fouling and cleaning in bipolar membrane electrodialysis of industrial glutamate production wastewater. Sep Purif Technol 79:103–113
4. He G, Li Z, Zhao J, Wang S, Wu H, Guiver MD et al (2015) Nanostructured ion-exchange membranes for fuel cells: recent advances and perspectives. Adv Mater 27:5280–5295
5. Varcoe JR, Slade RCT (2005) Prospects for alkaline anion-exchange membranes in low temperature fuel cells. Fuel Cells 5:187–200
6. Mizutani Y, Nishimura M (1970) Studies on ion-exchange membranes. XXXII. Heterogeneity in ion-exchange membranes. J Appl Polym Sci 14:1847–1856
7. Hale DK, McCauley DJ (1961) Structure and properties of heterogeneous cation-exchange membranes. Trans Faraday Soc 57:135–149
8. Mizutani Y, Kusumoto K, Nishimura M, Nishimura T (1975) Microheterogeneity of anion exchange membranes. J Appl Polym Sci 19:2537–2543
9. Kulshrestha V, Chatterjee U, Sharma S, Makwana BS, Maru PD (2015) Large scale preparation of polyethylene based ion exchange membranes and their application for water desalination. Macromole Sympo 357:194–199
10. Pozniak G, Trochimczuk W (1982) Chloromethylation of the polyethylene/poly(styrene-co-divinylbenzene) system. J Appl Polym Sci 27:1833–1838
11. Kang K, Kang PH, Nho YC (2006) Preparation and characterization of a proton-exchange membrane by the radiation grafting of styrene onto polytetrafluoroethylene films. J Appl Polym Sci 99:1415–1428

12. Lei Y-l, Luo Y-j, Chen F, Mei L-h (2014) Sulfonation process and desalination effect of polystyrene/PVDF semi-interpenetrating polymer network cation exchange membrane. Polymers 6:1914–1928
13. Ezzeldin HA, Apblett A, Foutch GL (2010) Synthesis and properties of anion exchangers derived from chloromethyl styrene codivinylbenzene and their use in water treatment. Int J Polym Sci 9
14. Pan Q, Liu S, Xie J, Jiang M (1999) Synthesis and characterization of block-graft copolymers composed of poly(styrene-b-ethylene-co-propylene) and poly(ethyl methylacrylate) by atom transfer radical polymerization. J Polym Sci, Part A: Polym Chem 37:2699–2702
15. Pantelić N, Andria SE, Heineman WR, Seliskar CJ (2009) Characterization of partially sulfonated polystyrene-block-poly(ethylene-ran-butylene)-block-polystyrene thin films for spectroelectrochemical sensing. Anal Chem 81:6756–6764
16. Sun Koo J, Kwak N-S, Hwang TS (2012) Synthesis and properties of an anion-exchange membrane based on vinylbenzyl chloride–styrene–ethyl methacrylate copolymers. J Membr Sci 423–424:293–301
17. Kumar V, Kumar P, Nandy A, Kundu PP (2015) Crosslinked inter penetrating network of sulfonated styrene and sulfonated PVdF-co-HFP as electrolytic membrane in a single chamber microbial fuel cell. RSC Adv 5:30758–30767
18. Chen Z, Huang D, Hwang J-Y (2019) Effect of styrene addition on chemically induced grafting of 4-vinylbenzyl chloride onto low-density polyethylene for anion exchange membrane preparation. Polym Int 68:972–978

Part III
Metallurgy

Pilot Plant Testing of Microwave/Plasma Pig Iron Nuggets and Syngas Productions

Xiaodi Huang, Jiann-Yang Hwang, and Rick Kauppila

Abstract This paper introduces a pilot plant test of pig iron nuggets and syngas productions using a microwave plasma rotary hearth furnace (MW/PA-RHF). In addition, the furnace is integrated with a bag house, a H2S wet scrubber, a syngas compressor, two syngas storage tanks, and a syngas monitoring system to process the produced syngas. Limited tests showed molten pig iron nuggets and syngas can be produced concurrently using this integrated system.

Keywords Steelmaking · Microwave · Plasma · Iron

Introduction

The project was to upgrade an existing 100 kg/hr microwave rotary hearth furnace (MW-RHF) with two additional 75 kW 915 MHz microwave generators to increase its capacity and then conduct testing and evaluation to optimize the process. The 100 kg/hr MW-RHF was used originally to produce direct reduced iron (DRI) as shown in Fig. 1.

In addition, the project developed pig iron nuggets and syngas co-production capability by adding: (1) a plasma heating system comprising of 52 electric arcing guns, 26 5-kW AC/DC power supplies, and a circulating water cooling system; (2) a bag house; (3) a H2S wet scrubber; (4) a syngas compressor; (5) two syngas storage tanks; and (6) a syngas monitoring system.

X. Huang (✉) · J.-Y. Hwang
Department of Materials Science and Engineering, Michigan Technological University, Houghton, MI, USA
e-mail: 18201560720@139.com

R. Kauppila
U.P. Steel, Houghton, MI, USA

B. Li et al. (eds.), *Materials Engineering—From Ideas to Practice: An EPD Symposium in Honor of Jiann-Yang Hwang*, The Minerals, Metals & Materials Series,
https://doi.org/10.1007/978-3-030-65241-8_6

Fig. 1 MW-RHF for DRI pilot plant testing. (Color figure online)

Experiments

Furnace Upgrade Design

Figure 2 shows a top inside view of the MW-RHF furnace. In operation, carbon particles are charged into the MW/RF-RHF furnace in a thin layer and a mixture of iron ore concentrate and pulverized coal was spread over it in a thick layer. Three processes happen during the MW/PA-RHF heating: iron ore partial reduction, iron nugget formation, and in situ syngas reforming.

Iron Ore Partial Reduction

First, microwave energy heats the layer of iron ore and coal mixture to approximately 800 °C in 10–20 min. During microwave heating, the coal serves as the reducing agent for iron oxide, an auxiliary heating source via an exothermal oxidation reaction, and the carbon and hydrogen source for the syngas generation. The iron oxide serves as the source of iron units for steel production as well as the source of oxygen for the

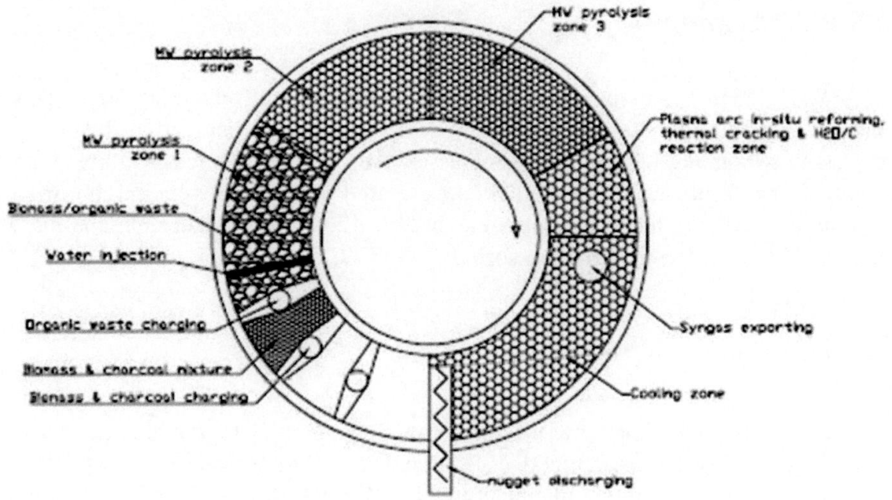

Fig. 2 MW/PA-RHF top inside view

carbon partial reaction to form CO. In comparison with conventional gasification, the oxygen comes from iron oxide (chemical looping) instead of from pure oxygen produced by an oxygen plant. During the microwave heating, iron ore is reduced into DRI. After DRI is formed, the feed material becomes a poor microwave absorber due to the formation of networked metallic iron. The microwave heating is designed to result in approximately 50–70% metallization with high heating efficiency. Volatiles in coal (primarily methane, CH_4) are also released during the process.

Iron Smelting

The iron and carbon content in the feedstock can be controlled to the Fe–C eutectic composition (4.26%C) through the feedstock recipe. At the eutectic, the melting point of iron is 1154 °C. In the MW/PA-RHF, the plasma arc heating takes over after the MW heating to complete the iron ore reduction and melt the eutectic Fe–C material into pig iron nuggets. The ash from coal and impurities in iron ore form slag. The slag composition can be adjusted by adding fluxing agents in the feedstock to form a slag suitable for desulphurization and dephosphorization with lower melting point, lower viscosity, proper basicity, and easy separation from the pig iron nuggets after cooling. The remaining underlying carbon layer functions as an isolator between the molten nuggets and slag and the refractory base in the furnace and facilitates discharging the produced nuggets and slag. The produced pig iron nuggets can be used as a feed material for ferrous foundries or made into steel.

In Situ Reforming

The MW/PA-RHF has a specially designed smelting/in situ reforming zone. This zone is constructed by lowering the ceiling of the furnace chamber for installing the plasma arcing and forcing better contact of the gases at high temperatures with the remaining carbon in the feedstock. At elevated temperatures and carbon-rich environment, iron ore reduction complete, hydrocarbons in off-gas crack, and most water and CO_2 react with carbon to form H_2 and CO.

Additional Microwave Power

There are totally three 75-kW 915-MHz microwave generators to provide microwave power for the furnace after adding two additional 75-kW microwave generators. The two microwave generators are manufactured by Microdry Inc.

Auxiliary Equipment

We planned an iron ore, coal, and limestone feeding system and selected Dynamic Air to design the details and manufacture the weighing, mixing, and feeding for the system. To build the capability of co-production of pig iron nuggets and syngas, we designed additionally (1) the electric arcing system comprised of 52 electric arcing guns, 26 5-kW AC/DC power suppliers, a circulated water cooling system, and a gas station; (2) the syngas processing system comprising of a bag house, a H_2S wet scrubber, a syngas compressor, two syngas storage tanks, and a syngas monitoring system. The syngas monitoring system includes a high concentration CO monitor, a CO_2 monitor, a multi-gas monitor, an electronic pressure, temperature, and flow rate meter, and four CO alarms (Fig. 3).

Testing

Two types of iron ores were tested which were donated by U.P. Fabrication and Superior Mineral Resources. The pulverized coal was purchased from DTE Petcoke LLC. Limestone was also donated by U.P. Fabrication. These feed materials were analyzed and processed to meet the test requirements.

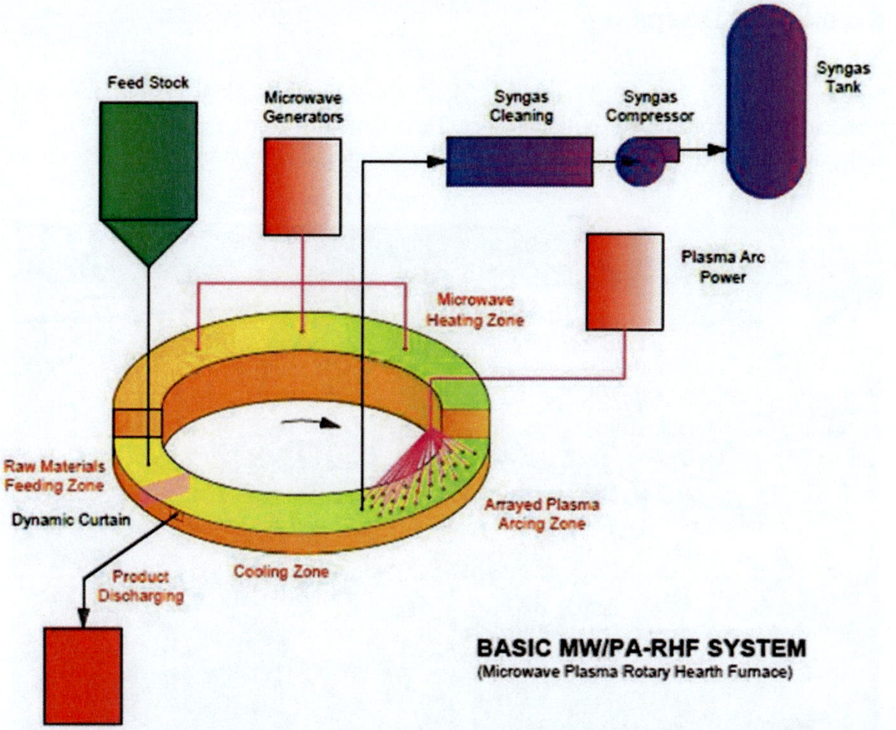

Fig. 3 MW/PA-RHF system for iron nuggets and syngas pilot plant testing. (Color figure online)

Furnace Operations

Variables such as material feed rate, rotation speed, microwave energy input, electric arcing energy, and operation time have been varied and tested. The temperatures at different locations of the furnace chamber were measured with optical pyrometers through multiple windows.

Results and Discussion

Figures 4, 5, 6, 7, 8, 9, 10, 11, 12, 13, 14, 15 show the pilot plant system designed, fabricated, and completed in this project. Tests have shown molten pig iron nuggets and syngas can be produced concurrently by this integrated system. Figure 16 shows

Fig. 4 A top view of MW/PA-RHF and material feeding system. (Color figure online)

Fig. 5 A view of microwave heating on iron ore through a chamber window. (Color figure online)

Fig. 6 Gas station to provide gas for the 52 electric arcing torches. (Color figure online)

Fig. 7 Explosion valve at 5 psi. (Color figure online)

the pig iron nugget produced. The highest CO+ combustible concentration of the furnace off-gas reached 70%.

Co-production of pig iron nuggets and syngas by MW/PA-RHF is a new development of microwave-assisted DRI making technology. It involves basic principle, process design, and facility development. We have demonstrated the technology's

Fig. 8 A view of the syngas treatment system. (Color figure online)

Fig. 9 Cooling water supply to the three microwave generators and 52 electric arcing torches. (Color figure online)

feasibility in the scale of pilot plant unit with associated facilities. It opens a door for further improvement and could promote the research in the area of iron and syngas co-production.

Fig. 10 The 26 electric arcing power supplies, 52 power and cooling water cables and cooling water discharge line. (Color figure online)

Fig. 11 The syngas treatment sytem comprising a bag house, a H_2S scrubber and a syngas compressor. (Color figure online)

Fig. 12 A view of the 52 electric arcing torches. (Color figure online)

Fig. 13 Firing the electric arc torches. (Color figure online)

Fig. 14 A view of the electric arc glaring. (Color figure online)

Fig. 15 Syngas storage tanks at 200 psi. (Color figure online)

Fig. 16 Iron nuggets produced using MA/PA-RHF. (Color figure online)

Acknowledgements This work is supported by the MPSC (Michigan Public Service Commission) program.

Control of Copper Loss in Flash Smelting Slag

Karen Cornejo, Mao Chen, and Baojun Zhao

Abstract Pyrometallurgical route is the most effective way to extract copper from Cu–Fe–S ores. In the smelting process, the concentrate is partially oxidized to Cu-rich matte and most of the impurities are reported to the slag. Copper content in the slag represents the direct recovery of copper in the smelting process. High-temperature experiments have been carried out to investigate the effects of slag composition, matte grade, and temperature on copper content in the smelting slag. Industrial and synthetic slag samples were equilibrated with matte at different temperatures in argon gas flow. The quenched samples were analysed by electron probe X-ray microanalysis to obtain the microstructure and phase compositions at high temperatures. Optimised parameters including slag composition, matte grade, and temperature have been proposed for the flash smelting furnace operations to minimize the copper content in the slag. The experimental results are also compared with the FactSage calculations.

Keywords Copper loss · Flash smelting · Slag · Liquidus temperature · Viscosity

Introduction

Copper is produced around the world, and South America plays an essential role in copper extraction with 40% mined in the world [1]. Approximately 80% of the copper is processed from Cu–Fe–S ores. The pyrometallurgical process is the most effective way to extract the copper from the Cu–Fe–S ores which includes smelting, converting, and refining steps [1]. The demand for copper is rising continuously and the amount and purity of the ores for its extraction have been decreased [2]. Continuous improvement of the extractive processes is essential to keep developing the products at high quality and high recovery. One of the main problems that have in the smelting plants is the metal losses in slags [2]. The understanding of the principles which are involved in the metal production and its losses will help the copper industry to optimise operational parameters and maximise the profits in its production.

K. Cornejo · M. Chen · B. Zhao (✉)
The University of Queensland, Brisbane, Australia
e-mail: baojun@uq.edu.au

B. Li et al. (eds.), *Materials Engineering—From Ideas to Practice: An EPD Symposium in Honor of Jiann-Yang Hwang*, The Minerals, Metals & Materials Series,
https://doi.org/10.1007/978-3-030-65241-8_7

Copper smelting slag is a mix of different oxides originated from the oxidation of the Cu–Fe–S ore and flux. These components include mainly ferrous oxide (FeO), ferric oxide (Fe_2O_3), and silica (SiO_2) with minor components alumina (Al_2O_3), calcium oxide (CaO), and magnesium oxide (MgO) [1]. In the smelting process, there are two types of copper losses. One is a loss by chemical dissolution and the other is mechanical entrainment of matte droplets [2]. Flash smelting furnace (FSF) is one of the major copper-making technologies used in Chile. It has been reported that chemical dissolution of copper in slags is the major loss of copper in the flash smelting slag [3]. Two possible types of chemical dissolution can affect the copper loss: oxidic and sulphidic copper dissolution [4]. The amount of copper corresponding to the oxidic copper in slags can be calculated as [4]

$$wtpctCu_{oxidic\,in\,slag} = A * a_{CuO_{0.5}} \tag{1}$$

$$Cu_l + \frac{1}{4}O_{2g} = CuO_{0.5} \tag{2}$$

where A is the constant of proportionality and activity of $CuO_{0.5}$ related to the equilibrium constant of Eq. (2). Values of the constant proportionality found in the literature have a range between 22.7 and 35.9, depending on the Fe/SiO_2 ratio, temperature, and CaO, Al_2O_3, and MgO contents [4].

Sulphidic copper dissolution had been investigated by a few researchers. The research made by Nagamori [5] was based on thermodynamic properties of matte and slag with the association of Cu–S bond to the copper dissolution. With this research, the sulphidic copper dissolution was established as a function of the percentage of sulphur in slag and the percentage of copper in the matte [4]. This methodology is useful in operating plants, due to S% in slag and Cu% in matte are known values. The amount of copper corresponding to the sulphidic copper in slags can be calculated as [4]

$$wtpctCu_{sulphidicinslag} = 0.00495 * wt\%S_{slag} * wt\%Cu_{matte} \tag{3}$$

where wt% S is the weight percentage of sulphur in slag and wt% Cu is the content of copper in the matte [4].

The copper loss into the slags has been a topic of interest for researchers and industrial operators. Understanding the effects of temperature, matte grade, and slag composition to reduce copper loss in the slags is essential to develop an optimal smelting process. These parameters will be studied in the conditions close to the operation in Chuquicamata Smelter to find an optimal range which reduces the copper loss in the slags.

The operational parameters of the flash smelting furnace in Chuquicamata Smelter have a range between 1230 and 1320 °C for slag temperature, from 1 to 1.56 for iron to silica ratio in the slag and 54–75 wt% of Cu for the matte grade [1]. Furthermore, the slag and matte used for this study were provided by Chuquicamata Smelter. The parameters to be studied for copper loss in the smelting slag included

(1) matte grade at 55, 60, and 65 wt% Cu.
(2) temperature at 1250, 1300, and 1300 °C.
(3) effect of slag composition: iron to silica ratios (Fe/SiO_2) of 1.2, 1.4, and 1.6 and calcium oxide (CaO) at 2, 4, and 6 wt%.

The experimental results will be compared with the calculations of FactSage.

Experimental

The experimental procedure includes preparation of samples, high-temperature treatment of these samples following by rapid quenching samples to water. The quenched samples were mounted, polished, and carbon-coated for microscopical analyses.

The initial copper matte and slag samples were collected from the flash furnace by Codelco Chuquicamata smelter. The compositions of the matte and slag analysed by XRF are shown in Tables 1 and 2.

The initial copper matte and slag were mixed with different chemicals to obtain the required compositions. FeS was added to prepare low-grade copper matte, SiO_2 was added to decrease the Fe/SiO_2 ratio of the slag, and the CaO from $CaCO_3$ was used to control its concentrations in the slag. Pure FeS, SiO_2, and $CaCO_3$ were supplied by Sigma-Aldrich.

One gram matte and 1 g slag were mixed in an agate mortar and pelletized. The pellet was placed in an alumina crucible (7 mm ID and 30 mm height). The experiments were performed in a vertical furnace with an alumina reaction tube (30 mm ID). The temperature inside the reaction tube was measured by a working thermocouple next to the sample. The sealed reaction tube was flashed by ultra-high-purity Ar. Preliminary experiments at 1350 °C for 30, 60, and 120 min shown that the compositions of the slags did not change significantly indicating the dissolution of copper from the matte and alumina from the crucible reached equilibrium. The processing time was chosen to be 60 min for all experiments.

After equilibration, the samples were dropped into water for rapid quenching. The dried samples were mounted in epoxy resin and polished for EPMA analysis. A JXA-8200 electron probe X-ray microanalyser (Japan Electron Optics Ltd.) was used for microstructural and compositional analyses. An accelerating voltage of

Table 1 Bulk composition of the matte from XRF analysis (wt%)

Cu	Fe	S	Al	As	Ca	Mg	Pb	Si	Zn
67.1	10.8	20.4	0.02	0.79	0.01	0.02	0.13	0.07	0.71

Table 2 Bulk composition of the slag from XRF analysis (wt%)

Cu_2O	FeO	CaO	SiO_2	Al_2O_3	MgO	S	ZnO	As_2O_3	PbO	Fe/SiO_2
2.2	58.5	0.8	28.3	6.2	0.3	0.5	2.6	0.5	0.1	1.61

15 kV and a probe current of 15 nA were used. ZAF correction procedure was applied. The standards used for analysis were from Charles M. Taylor Co. (Stanford, California): Al_2O_3 for Al, $CaSiO_3$ for Ca, and Si, Fe_2O_3 for Fe, and from Micro-Analysis Consultants (Cambridge): Cu_2O for Cu. The average accuracy of the EPMA measurements was estimated to be within ± 1 wt%. EPMA analysis can only analyse the elemental composition but cannot determine the electronic state of the elements. Thus, iron is calculated as FeO and copper as Cu_2O for the presentation purpose. The predictions were calculated by FactSage 7.3 with the module "equilib", the databases "FactPS", "FToxide", and "FTmisc" [6]. The SO_2 partial pressure was fixed at 0.2 atm and the O_2 partial pressure was fixed at 10^{-7}–10^{-8} atm depending on the matte grade and temperature for the calculations.

Results and Discussion

Typical microstructures of the samples quenched from 1250 °C are shown in Fig. 1. It can be seen from the figures that the slag was fully liquid at 1250 °C. Small amounts of matte droplets may be present inside the slag which confirmed that most of the copper loss in the slag is in form of chemical dissolution. The focus of this study is therefore the discussion on effects of operating parameters on the chemically dissolved copper in the smelting slag. As EPMA can only analyse the elemental concentration, the chemically dissolved copper in the slag is shown as "Cu_2O" which includes copper oxide and copper sulphide. The compositions of the liquid slag and the matte in the samples measured by EPMA are given in Tables 3 and 4, respectively.

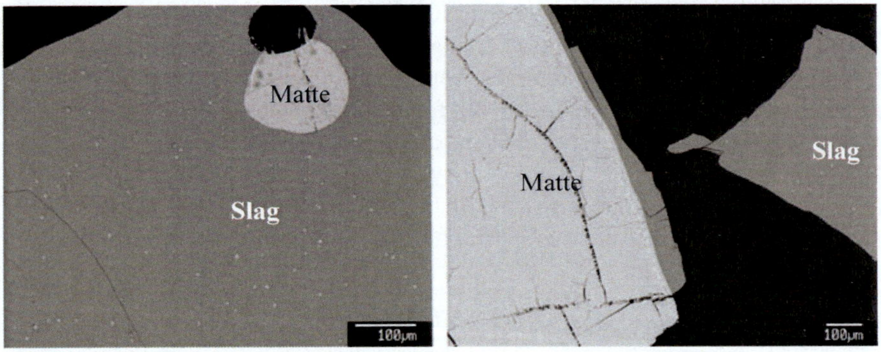

Sample 7, $Fe/SiO_2 = 1.6$ Sample 9, $Fe/SiO_2 = 1.4$

Fig. 1 Typical microstructures of the samples quenched from 1250 °C

Table 3 Compositions of the liquid slag in the samples quenched from 1250 °C

Sample	"FeO"	SiO_2	"Cu_2O"	CaO	Al_2O_3	MgO	S	Fe/SiO_2
7	59.7	28.6	0.3	1.8	8.7	0.3	0.7	1.62
9	55.9	30.8	0.2	2.2	9.7	0.3	1.0	1.41

Table 4 Compositions of the matte in the samples quenched from 1250 °C

Sample	Fe	Cu	S	As
7	9.5	67.4	22.7	0.4
9	14.6	60.5	24.3	0.5

Effect of Iron to Silica Ratio on Dissolved Copper in Smelting Slag

The samples with different iron to silica ratios (Fe/SiO_2) were treated at 1250 °C for 60 min in equilibrium with the fixed matte grade. The experimental results are also compared with FactSage predictions. It can be seen from Fig. 2 that dissolved copper in slag increases with increasing Fe/SiO_2 ratio. The slag in equilibrium with higher grade matte contains higher dissolved copper. The same trends are obtained in both experimental results and the FactSage calculations. However, it can be seen that the predicted values by FactSage are much higher than the experimental results. Part of the reasons is that the SO_2 and oxygen partial pressures were not controlled in the high-temperature experiments.

The iron to silica ratio (Fe/SiO_2) in the smelting slag is usually controlled by the addition of silica flux. To decrease the dissolved copper in the slag, more silica is required. Increased silica in the slag also decreases the liquidus temperature of the slag [3]. However, a number of disadvantages are associated with the low Fe/SiO_2: (1) increased cost of the flux; (2) increased liquid slag viscosity; (3) increased slag

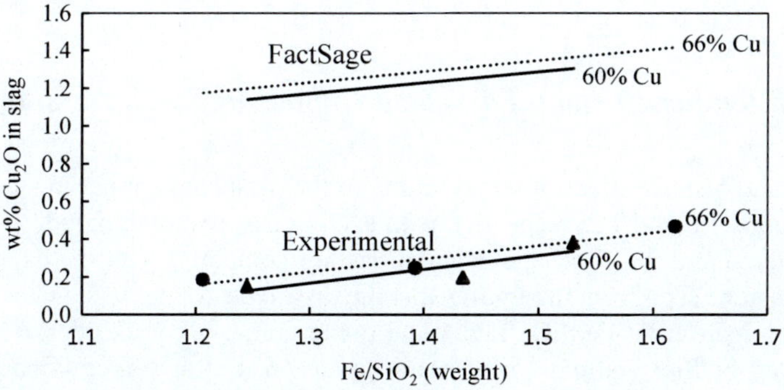

Fig. 2 Effect of Fe/SiO_2 ratio on the dissolved copper in slag at 1300 °C

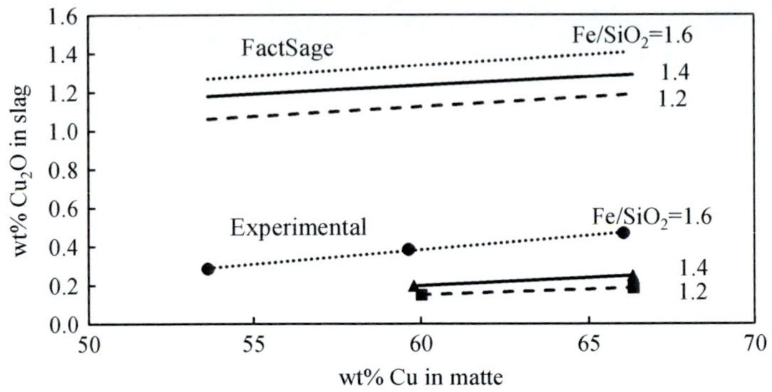

Fig. 3 Effect of matte grade on dissolved copper in slag at 1300 °C

volume which will increase the heat loss and the cost for slag treatment. Thus, an optimum Fe/SiO_2 needs to be determined according to the maximum economic benefits.

Effect of Matte Grade on Dissolved Copper in Smelting Slag

Figure 3 shows the dissolved copper in slag as a function of matte grade at Fe/SiO_2 of 1.2, 1.4, and 1.6. It can be seen in Fig. 3 that at the same Fe/SiO_2, the dissolved copper in slag increases with increasing matte grade. FactSage calculations show the same trend but the predicted copper solubilities are much higher. Higher grade matte means that more iron has been oxidised from the sulphide concentrate resulting in a higher Fe/SiO_2 in the slag. Furthermore, high matte grade is always associated with the high Fe/SiO_2, both can increase the dissolved copper in slag. On the other hand, low-grade matte will reduce the productivity of the smelter. It is desirable to maintain a balance between the matte grade and Fe/SiO_2 in the slag.

Effect of Temperature on Dissolved Copper in Smelting Slag

Figures 4 and 5 show effect of temperature on the dissolved copper in slag at fixed Fe/SiO_2 ratios 1.6 and 1.2, respectively. In each figure, two matte grades 60 and 66 are considered and the FactSage calculations are compared with the experimental results. It can be seen from the figures that the dissolved copper in slag has different behaviors for different Fe/SiO_2 ratios and matte grade. It can be seen from Fig. 4 that at fixed Fe/SiO_2 ratio 1.6 and matte grade 66, the dissolved copper in slag increases with increasing temperature. However, if 60 wt% Cu matte is produced, it seems that the solubility of copper in slag decreases slightly when the temperature

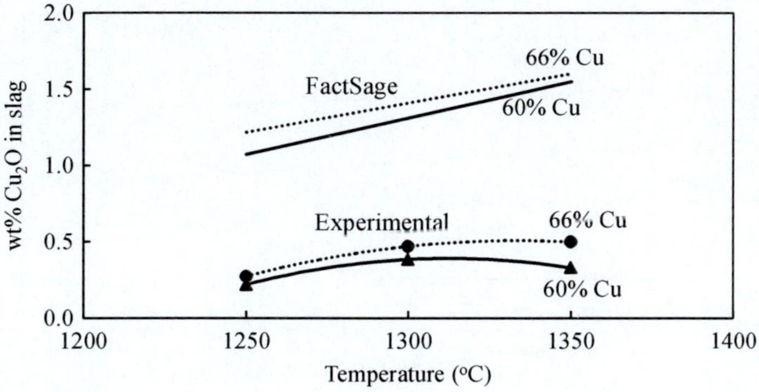

Fig. 4 Effect of temperature on dissolved copper in slag at Fe/SiO_2 1.6

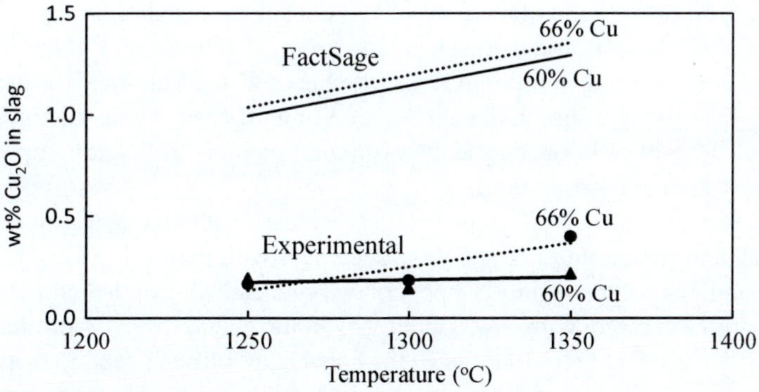

Fig. 5 Effect of temperature on dissolved copper in slag at Fe/SiO_2 1.2

increases from 1300 to 1350 °C. It can be seen from Fig. 5 that, at Fe/SiO_2 1.2 and matte grade 66 wt% Cu, the dissolved copper in slag increases with increasing temperature. However, if 60 wt% Cu matte is produced, the copper solubility in slag is almost independent of the temperature. It seems that low-temperature and low-grade matte operation can reduce the copper solubility in the smelting slag. Figures 4 and 5 also show that the FactSage predictions are much higher than the experimentally determined results. Less sensitivity of the copper solubility to the temperature is not predicted by the FactSage.

Effect of Calcium Oxide on Dissolved Copper in Smelting Slag

Experiments were performed at 1300 °C with the fixed matte grade 60 wt% Cu and iron to silica ratios 1.2 and 1.6. FactSage calculations were conducted under the same conditions.

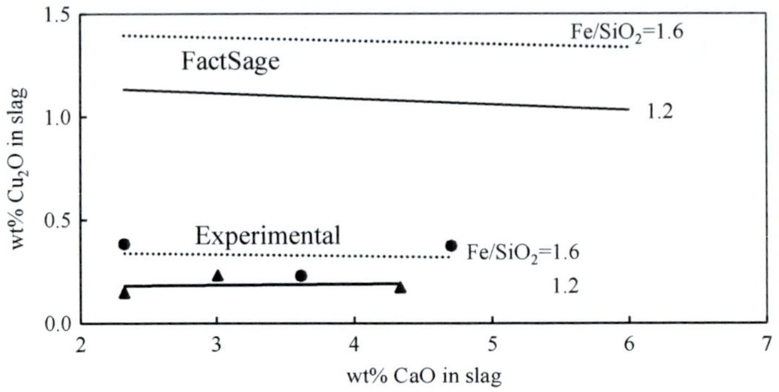

Fig. 6 Effect of CaO on dissolved copper in slag at 1300 °C

It can be seen from Fig. 6 that the dissolved copper in slag slightly decreases with increasing CaO concentration in the slag. However, the addition of CaO will increase the material cost, liquidus temperature of the slag and slag volume. It is usually not practical to use CaO as a flux in the copper smelting process. Again the figure shows a significant difference between the experimental results and FactSage predictions although they have the same trend.

In conclusion, the dissolved copper in slag increases with increasing Fe/SiO_2 ratio, matte grade, and temperature but slightly decreases with increasing CaO concentration in the slag. Use of low-grade copper concentrate can control the temperature and produce low-grade matte with low copper loss in the slag. However, productivity of the smelting furnace and requirement of the converting furnace need to be taken into account. Mechanically entrained matte droplets also need to be controlled together with the chemically dissolved copper. Optimised Fe/SiO_2 ratio, matte grade, and temperature are different in different smelters. The results were based on the flash smelting slag but they can also be applied to other smelting furnaces.

Conclusions

This research explored the effects of different parameters used in copper smelter on the dissolved copper in the flash smelting slag. The parameters include matte grade, temperature, and slag composition. FactSage calculations were also compared to the experimental results.

The experimental results indicated that a lower temperature can reduce the dissolved copper in the slag. However, lower operating temperatures will increase the viscosity of the slag which may increase the matte droplet in the slag. Iron to silica ratio can increase the dissolved copper in the slag. The addition of more silica can reduce the dissolved copper in the slag but will increase the viscosity of the slag. The copper solubility in the slag increases with increasing the matte grade. The matte grade produced in the smelting furnace also needs to consider the requirement of the converting furnace and overall productivity.

The FactSage predictions show the same trends as the experimental results. However, the calculated data are generally higher than the experimental results. The reasons could be the different conditions used in the calculations and experiments. The FactSage databases also need to be further optimised.

References

1. Schlesinger ME, King MJ, Sole KC, Davenport WG (2011) Extractive metallurgy of copper, 5th edn. Elsevier, Oxford
2. Bellemans I, De Wilde E, Moelans N, Verbeken K (2018) Metal losses in pyrometallurgical operations—a review. Adv Colloid Interface Sci 255:47–63
3. Chen M, Contreras L, Zhao B (2016) Comparison of slag chemistry between teniente converter and flash smelting furnace. In: Proceedings of 9th international copper conference, Kobe, Japan, November 13–16, 2016, pp 967–975, The Mining and Materials Processing Institute of Japan (MMIJ) and Japan Mining Industry Association (JMIA)
4. Mackey PJ (1982) The physical chemistry of copper smelting slags—a review. Can Metall Q 21:221–260
5. Nagamori M (1974) Metal loss to slag: part II. Oxidic dissolution of nickel in fayalite slag and thermodynamics of continuous converting of nickel-copper matte. Metall Trans 5:539–548
6. Bale CW, Bélisle E, Chartrand P, Decterov SA, Eriksson G, Gheribi AE, Hack K, Jung IH, Kang YB, Melançon J, Pelton AD (2016) FactSage thermochemical software and databases, 2010–2016. Calphad 55: 1–19

Effect of Boron Iron Concentrate on the Strength of Preheated Iron Ore Pellets

Li Ma, Gele Qing, Zhixing Zhao, and Baojun Zhao

Abstract Pellet is one of the important feeds to iron blast furnace and increased proportion of the pellet is used around the world. Pellet production is cleaner than sinter and its uniform properties allow stable blast furnace operations. Boron iron concentrate is a by-product during production of boron and contains approximately 54% Fe, 3–6% B_2O_3, and 10% MgO. B_2O_3 can form liquid binding phase inside the pellet at relatively lower temperature which reduces the fuel and refractory consumption during the pellet production. MgO can reduce the reduction expansion rate of the pellets. Effects of boron iron concentrate, preheating temperature, and time on the strength of the preheated pellet have been investigated. It was found that the compressive strength of the preheated pellet with addition of boron iron concentrate was increased with increasing preheating temperature and time. When the pellets were preheated at 1030 °C for 25 min, the strength of the pellets with 50% boron iron concentrate can reach more than 600 N/P. The optimized parameters can significantly increase the strength of the pellet and reduce the loop formation of the rotary kiln.

Keywords Boron iron concentrate · Magnetite concentrate · Pellet · Preheating strength

---

L. Ma · G. Qing (✉) · Z. Zhao
Shougang Research Institute of Technology, Beijing, China
e-mail: baojun@uq.edu.au

B. Zhao
The University of Queensland, Brisbane, Australia

L. Ma · G. Qing · Z. Zhao
Beijing Key Laboratory of Green Recyclable Process for Iron and steel Production Technology, Beijing, China

B. Li et al. (eds.), *Materials Engineering—From Ideas to Practice: An EPD Symposium in Honor of Jiann-Yang Hwang*, The Minerals, Metals & Materials Series,
https://doi.org/10.1007/978-3-030-65241-8_8

Introduction

China has rich resources of boron iron ore. The boron ore is mainly used to produce boric acid, borax, and other compounds of boron. During the processing of boron ore, a large amount of industrial by-products, boron-containing iron ore, or boron iron concentrate are generated after the magnetic separation from boron concentrate. Boron iron concentrate contains approximately 54% Fe, 3–6% B_2O_3, and 10% MgO [1].

It has been reported that the magnesium in the boron iron concentrate is beneficial to reduce the reduction expansion rate of the pellets and improve the metallurgical properties of the pellets [2–5]. On the other hand, the boron in the concentrate is easy to form liquid phase inside the pellet at relatively lower temperature which can lower the roasting temperature of the pellets. Therefore, it is of great significance to explore the use of boron iron concentrate for efficient utilization of boron-containing resources and improvement of the metallurgical properties of pellets. The main process for producing pellets in China is the grate-rotary kiln process which includes drying, preheating, roasting, soaking, and cooling. The preheating is usually carried out in a rotary kiln. The preheating strength during the pellet preparation process is very important to the loop formation of the rotary kiln and stable production. This study focused on the effect of boron iron concentrate on the preheating strength of pellets.

Raw Material Properties and Research Methods

Raw Material Properties

This study used two types of iron-containing raw materials, magnetite concentrate, and boron iron concentrate. Their chemical compositions together with the bentonite binder are shown in Table 1. It can be seen from Table 1 that compared to the magnetite concentrate; the iron grade of the boron iron concentrate is 15% lower. However, the MgO content is 10.75% and the B_2O_3 content is 5.5% in the boron iron concentrate. The particle size distribution of the two iron concentrates is shown in Table 2, and the micro-morphologies of the mineral particles are shown in Fig. 1. It can be seen from Table 2 that more than 80% of both concentrates are small than 74 μm. The

Table 1 Chemical compositions of raw materials (wt%)

Material	T_{Fe}	FeO	CaO	MgO	SiO_2	Al_2O_3	B_2O_3	S	Ignition loss
Magnetite	68.0	28.5	0.43	0.51	3.89	0.46	–	0.05	−2.57
Boron iron	52.2	21.8	0.42	10.75	5.29	0.20	5.50	0.69	1.79
Bentonite	–	–	2.88	3.24	61.95	13.74	–	1.20	10.65

Table 2 Distribution of two iron concentrates (wt%)

Concentrate	+180 µm	180 ~ 120 µm	120 ~ 109 µm	109 ~ 96 µm	96 ~ 80 µm	80 ~ 74 µm	−74 µm
Magnetite	0.96	2.81	1.02	1.24	3.76	6.42	83.79
Boron iron	0.68	2.63	3.23	2.65	4.68	5.57	80.56

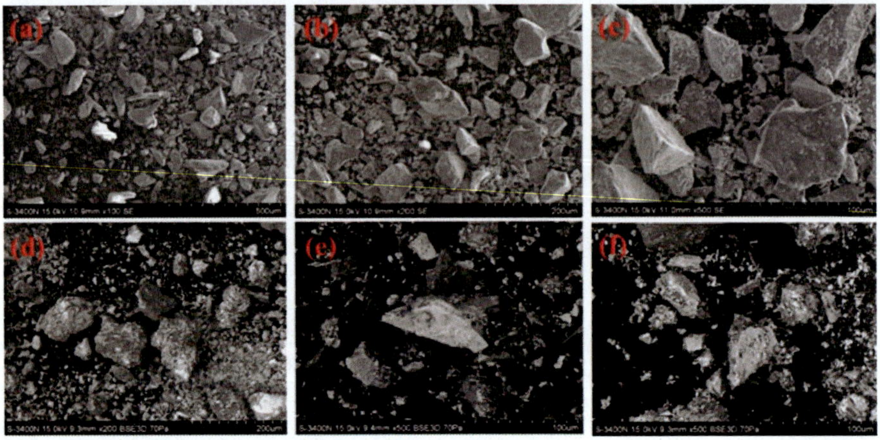

Fig. 1 Micro-morphologies of the iron concentrates used for pelletizing (**a–c**) magnetite concentrate; (**d–f**) boron iron concentrate. (Color figure online)

Table 3 XRD results showing the phases in boron iron concentrate (%)

Phase	Fe_3O_4	$MgBO_2(OH)$	CaB_2O_4
wt%	76.3	12	11.7

boron iron concentrate is slightly larger than the magnetite concentrate. Figure 1 shows that the particle surface of the boron iron concentrate is relatively rough, which is beneficial to the improvement of the subsequent pelletizing performance. X-ray powder diffraction (XRD) was used to analyze the phase composition of the boron iron concentrate. It can be seen from Table 3 that the boron iron concentrate consists of 76.3% magnetite, 12% szaibelyite ($MgBO_2(OH)$), and 11.7% calcium borate (CaB_2O_4). The bentonite used in the test is ordinary sodium bentonite with a colloidal value of 375 ml/15 g and a water absorption rate of 380% within 2 h.

A Thermogravimetry and differential thermal analyzer (TG-DTA) was used to analyze the weight loss behavior of the boron iron concentrate. A weight gain between 300 and 580 °C indicated the oxidation of the magnetite (Fe_3O_4) to hematite (Fe_2O_3). From 600 to 750 °C, a significant weight loss is attributed to the dehydration weight loss of the szaibelyite ($MgBO_2(OH)$) according to the following reaction [6]:

$$\left(MgBO_2(OH)\right) \rightarrow B_2O_3(s) + MgO(s) + H_2O(g) \tag{1}$$

Research Methods

The raw materials were weighed and mixed according to the plan. The pelletizing equipment was a disc pelletizing machine with a diameter of 800 mm. During the experiment, the rotation speed was controlled at 19 r/min and the inclination angle was 45°. The pelletizing time was 9 min. After the pelletizing experiment, the green pellets with a diameter of 10 ~ 12.5 mm were screened out for preheating. The preheating experiment was carried out in a tubular electric furnace. The compressive strength of the pellets was measured by a commercial compressive strength tester, and the microstructure was analyzed by a scanning electron microscope.

Results and Discussion

Experimental Results of Pelletizing and Preheating

Seven pelletizing experiments were carried out with different proportions of boron iron concentrate. The proportions of the boron iron concentrate were 0%, 10%, 20%, 40%, 50%, 80%, and 100% respectively, and the bentonite was fixed at 2.0%. The average drop strength of all green pellet was more than 5 times/0.5 m, and the compressive strength was higher than 8 N/P. It shows that the pelletizing performance and index of boron-containing green pellet are relatively good, and the boron iron concentrate can be used with other iron concentrates.

The pellets were preheated at three different maximum temperatures of 930, 980, and 1030 °C, respectively. The preheating time at the preheating temperatures of 930 and 980 °C was 22 min. The preheating time at 1030 °C was extended to 25 min. The preheated pellets were cooled for compressive strength test and the results are shown in Fig. 2. It can be seen from Fig. 2 that after preheating at 930 °C for 22 min, the strength of the preheated pellet gradually decreases with the increase of the ratio of boron iron concentrate. Up to 20% addition of the boron iron concentrate only slightly decreases the strength. More than 40% boron iron concentrate can significantly decrease the strength. For example, the strength of the preheated pellets without boron iron concentrate is 680 N/P. When the ratio of boron iron concentrate is increased to 50%, the strength of the preheated pellet is reduced to 512 N/P. Addition of more than 80% boron iron concentrate will decrease the compressive strength of the preheated pellet below 400 N/P. It indicates that with the conditions of preheating at 930 °C for 22 min, addition of the boron iron concentrate cannot make high-strength pellet. After the preheating temperature was increased to 980 °C, it can be seen from Fig. 2 that the strength of the preheated pellet with the addition of boron iron concentrate is generally higher than that preheated at 930 °C. However, the strength of the preheated pellet still decreases with increasing the proportion of the boron iron concentrate.

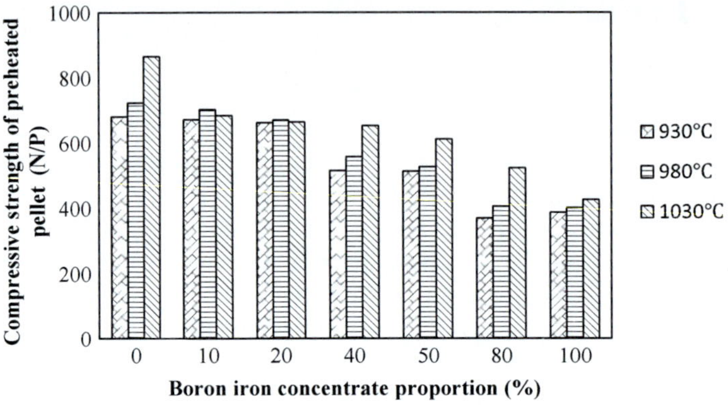

Fig. 2 Relationship between compressive strength of preheated pellet and proportion of boron iron concentrate at different temperatures

When the preheating temperature was further increased to 1030 °C and the preheating time was extended from 22 to 25 min, the strength of the preheated pellet was significantly increased. The compressive strength of the pellets with 50% boron iron concentrate can reach 600 N/P which can meet the production requirements.

The results above show that in the production of pellet by chain grate machine-rotary kiln process, the temperature and the time of the preheating need to be increased if boron iron concentrate is added. Enough temperature and time are required to decompose the B_2O_3-containing phases. It is beneficial to improve the strength of the preheating pellet and reduce the ring formation of the rotary kiln.

Influence of the Boron Iron Concentrate Addition on the Strength of the Preheated Pellets

The influence of the iron boron concentrate on the microstructure of the preheated pellets is shown in Fig. 3. The proportion of the iron boron concentrate is 0%, 50%, and 100%, respectively and the preheating temperature and time were 1030 °C and 25 min, respectively. It can be seen from the figure that the inside of the preheated pellet has a layered structure. As the proportion of the boron iron concentrate increases, the connections between the hematite particles are becoming weaker. This explained that the compressive strength of the preheated pellet decreased with increasing the proportion of the boron iron concentrate.

The image recognition method was used to identify and analyze the pore structure of the preheated pellets with different ratios of boron iron concentrates. Photoshop software was used to analyze the pores through different mineral brightness for the same magnification SEM image. The quantitative characterization of the pore structure is shown in Fig. 4 for three preheated pellets from 1030 °C. Figure 4a–c

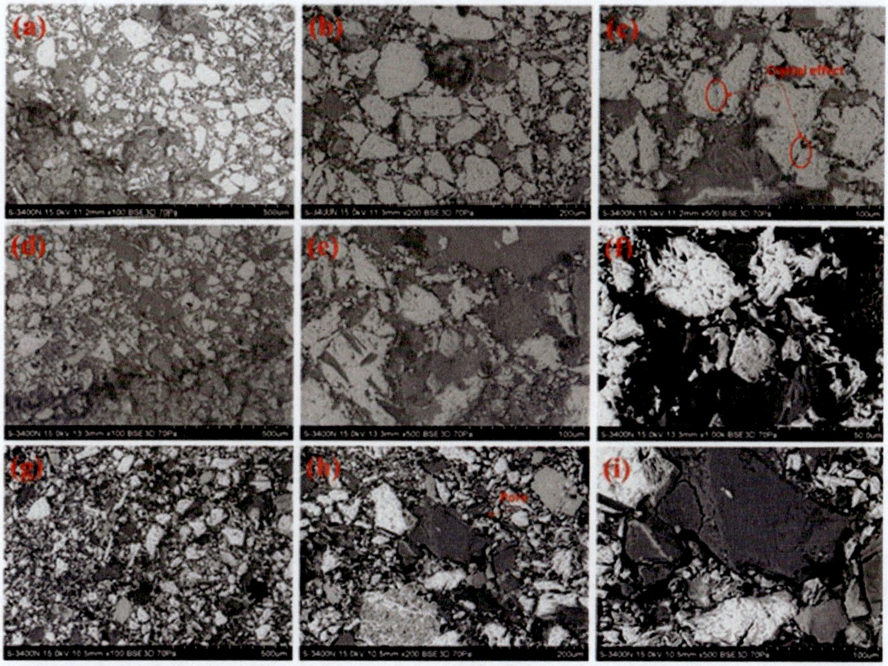

Fig. 3 Microstructures of the preheated pellets at different magnifications, (**a**–**c**) 0% boron iron concentrate; (**d**–**f**) 50% boron iron concentrate; (**g**–**i**) 100% boron iron concentrate. (Color figure online)

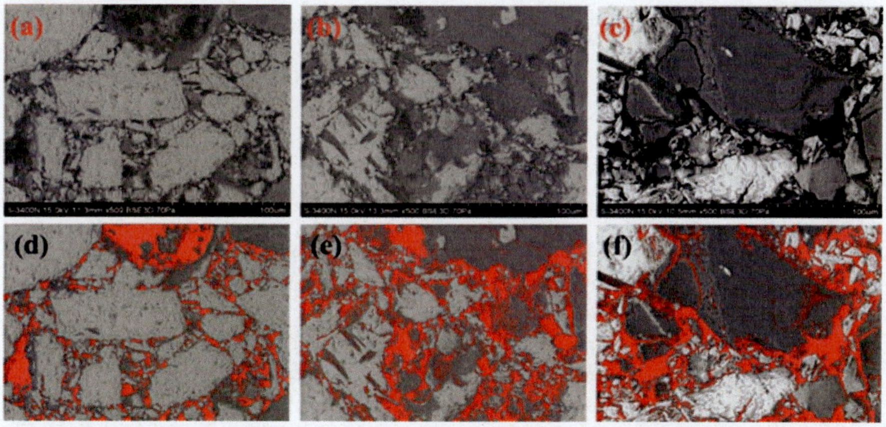

Fig. 4 Quantitative characterization of the pore structure of the preheated pellets under 500 magnification, (**a** and **d**) 0% boron iron concentrate; (**b** and **e**) 50% boron iron concentrate; (**c** and **f**) 100% boron iron concentrate. (Color figure online)

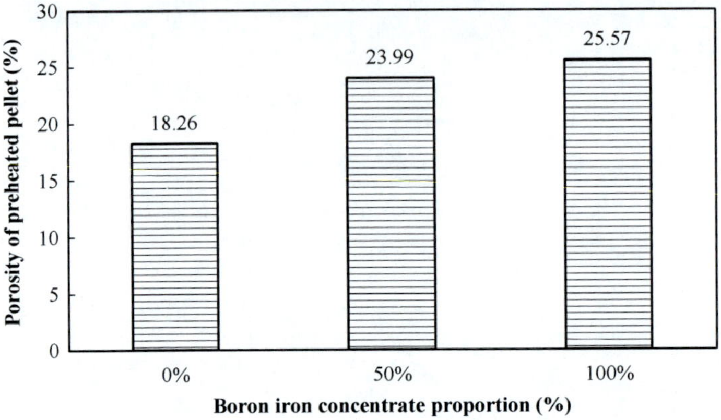

Fig. 5 Porosity of the preheated pellet with different proportions of the boron iron concentrate, 1030°C 25 min

is original SEM pictures and Fig. 4d–f is the images after the pores are filled. The porosities obtained based on the image analysis are 18%, 24%, and 26%, respectively, for the preheated pellets with 0%, 50%, and 100% boron iron concentrate as shown in Fig. 5.

It can be seen from Fig. 5 that the porosity of the preheated pellets gradually increases with the increase of the boron iron concentrate. The reason is that decomposition of szaibelyite ($MgBO_2(OH)$) resulted in the removal of crystal water and increased the pores of the remaining mineral at high temperature. At the same time, due to the decomposing endothermic heat of the ball, the heat input for the pellet preheating was insufficient. The increased ratio of the boron iron concentrate reduced the iron grade of the pellet, and the amount of iron oxide required for continuous crystal decreased. In conclusion, the increase in the porosity of the preheated pellet has caused a decrease in the compressive strength when born iron concentrate is used.

Conclusions

(1) Relatively high levels of MgO and B_2O_3, and rough surface of the boron iron concentrate are beneficial to pelletizing and pellet production.
(2) The weight gain of the boron iron concentrate between 300 and 580 °C is the result of magnetite oxidation. A significant weight loss between 600 and 750 °C is mainly caused by szaibelyite dehydration.
(3) With the increase of the proportion of boron iron concentrate, the compressive strength of the preheated pellet decreases, which is mainly caused by the increase of the porosity of the pellets; increasing the preheating temperature and prolonging the preheating time can the compressive strength of the preheated

pellets. According to laboratory test results, the preheating temperature 1030 °C, and the preheating time 25 min can achieve the compressive strength of the preheated pellets to be 600 N/P or more with 50% boron iron concentrate addition.

(4) The main components of the boron iron concentrate, iron oxide, MgO, and B_2O_3 are all beneficial to the production of iron pellet which promotes the efficient utilization of this by-product.

Acknowledgements This study was supported by National Key R&D Program of China (2017YFB0304300 and 2017YFB0304302).

References

1. Wang L (2019) Development suggestions and current situation of boron industry in Liaoning Province. Liaoning Chem Indus 48(7):680–682
2. Zhu D, Zhou W, Pan J, Chen D (2014) Improving pelletization of Brazilian hematite by adding boron-containing magnetite. J Central South Univ 45(2):348–355
3. Fu J, Bai G, Li G, Yang Y, Chen Y, Wang B, Zheng L, Yang A (2004) Study on the pellets of magnetite concentrate mixed with boron iron concentrate. Pelletizing Technol 1:2–6
4. Wang G, Xue Q, Wang J (2016) Strengthening the reduction process of carbon-containing pellet of boron-bearing iron concentrate. Sintering Pelletizing 41(6):31–37
5. Tian Y, Qin G, Zhang Y, Zhao L, Yang T (2020) Experimental research on pellet production with boron-containing concentrate, characterization of minerals, metals, and materials 2020. Miner Metals Mater Ser 91–102. https://doi.org/10.1007/978-3-030-36628-5_9
6. Jiao S, Jiang Z, Bu J (2010) Reaction synthesis and oxidation behaviours of Ludwigite. Adv Mater Res 146–147:475–480

Experimental Research on the Roasting of Brazilian Iron Concentrate in a Tube Furnace

Liu Lin and Zhao Qiang

Abstract In this paper, the method of preparing roasted pellets in a tube furnace is studied. In order to meet the performance standards of blast furnace ironmaking finished pellets, the compressive strength of the pellets must be above 2500 N/P. Hematite and magnetite were used as the main raw materials. The effects of preheating temperature, preheating time, roasting temperature, roasting time and the ratio of magnetite concentrate on the strength of pellets were studied. Under the conditions of preheating temperature of 975 °C, preheating time of 15 min, roasting temperature of 1300 °C, roasting time of 15 min, and the proportion of magnetite concentrate of 20–30%, the compressive strength 4948.3 N/P of roasted pellets was reached, thus meeting the performance indicators.

Keywords Roasted pellet · Magnetite · Hematite · Roasting time · Compressive strength

Introduction

China's steel output has been ranked first in the world in recent years [1]. As the world's magnetite reserves are decreasing and prices are increasing, more and more manufacturers use hematite as a raw material. Hematite has the advantages of low price, high iron content, and low impurity content [2], but the pellets produced from hematite have lower strength and poor ball formation [3–7]. Therefore, the research on the preparation of oxide pellets using hematite as a raw material has become a research subject [8–13].

A large number of studies have been conducted on the preparation of oxidized pellets in hematite at home and abroad. It has been pointed out in the literature [14] that it is necessary to increase the preheating and roasting temperatures of the pellets, while extending the preheating and roasting time. Some literature [15] also pointed out that by adding CaO, the basicity of the pellets can be increased, and

L. Lin · Z. Qiang (✉)
Zhongye Changtian International Engineering Co., Ltd., Changsha 410205, China
e-mail: 13319532520@163.com

B. Li et al. (eds.), *Materials Engineering—From Ideas to Practice: An EPD Symposium in Honor of Jiann-Yang Hwang*, The Minerals, Metals & Materials Series,
https://doi.org/10.1007/978-3-030-65241-8_9

the compressive strength of the roasted balls is in the range of 0.4–0.6. But below this range, the roasted ball strength will be significantly reduced. At the same time, adding too much CaO will significantly reduce the grade of TFe(total ferrum), which is contrary to the concentrate policy advocated by blast furnace iron making.

In this paper, Brazilian MBR hematite and domestic magnetic concentrate, were used as the main raw material. The effects of preheating temperature, preheating time, roasting temperature, roasting time and the ratio of magnetic concentrate on pellet strength were studied. Under the optimized experimental conditions, the finished pellets with good compressive strength and metallurgical properties were obtained, meeting the requirements of raw materials for the blast furnace iron making.

Raw Material Performance

Iron Concentrate

There are three kinds of iron concentrates used in this experiment: Brazilian MBR hematite and two kinds of domestic magnetite concentrates.

The chemical composition of MBR hematite and magnet concentrate is shown in Table 1. The physical properties of iron concentrate and the particle size composition of MBR after grinding and magnetite concentrate are shown in Tables 2 and 3, respectively.

As can be seen from Table 1, the FeO content of MBR hematite is only 0.19%, the iron grade is high (66.79%), the silicon content is low, and the content of other impurities such as P, S, and non-ferrous metal elements is very low. The particle size is fine, −0.074 mm reaches 91.14%, which is a relatively suitable raw material for pelletizing.

The FeO content of both kinds of magnetite concentrates is above 24%, and w(TFe)/w (FeO) is less than 2.7, which is the primary magnetite concentrates. As the main raw material of pellet, magnetite concentrate is conducive to the consolidation of pellet. When magnetite is oxidized to hematite, it is accompanied by structural changes, while lattice changes in the oxidation process and atoms on the surface of new crystals have high migration ability, which is conducive to the formation of crystal bonds between adjacent particles and the improvement of pellet strength. In addition, the oxidation of magnetite concentrate to hematite is an exothermic reaction, and the heat released by it is almost half of the totál heat consumption of the roasting process, which helps to make up for the lack of internal heat in the roasted hematite pellets. However, the oxidation of magnetite in pellets must be ensured when magnetite is used, If pellets are not oxidized sufficiently in preheating and high-temperature roasting, pellets would cause double-layer structure. Additionally in the high-temperature roasting zone, it would react with gangue SiO_2 to form a low-melting point compound. This will result in the formation of a liquid slag phase inside the pellet. Some concentric cracks also occur during cooling and shrinking.

Table 1 Chemical composition of iron concentrate/%

Mineral	TFe	FeO	SiO_2	Al_2O_3	CaO	MgO	K_2O	Na_2O	ZnO	PbO	CuO	S	P	LOI
MBR	66.79	0.19	2.47	0.82	0.10	0.038	0.071	0.014	0.014	0.017	0.0043	0.017	0.024	0.64
Magnetite 1	66.33	28.28	2.62	2.01	1.44	0.83	0.097	0.070	0.0016	0.0001	0.013	0.30	0.022	1.03
Magnetite 2	64.39	25.13	4.31	0.86	1.86	0.78	0.013	0.015	0.039	0.0097	0.0015	0.066	0.007	0.77

Table 2 Some physical properties of iron concentrate

Mineral	Bulk density/g cm^{-3}	True density/g cm^{-3}	Porosity/%	Static accumulation angle/°	Static sphericity index	Specific surface area/cm^2 g^{-1}
MBR	2.1719	4.55	52.27	34.34	0.19	520
Magnetite 1	2.2460	4.57	50.85	35.71	0.242	1650
Magnetite 2	1.862	4.71	60.47	33.28	0.651	1875

Table 3 Iron concentrate particle size composition/%

Granularity/mm	+ 0.150	−0.150 + 0.074	−0.074 + 0.043	−0.043
MBR	1.63	7.23	29.55	61.59
Magnetite 1	0.30	8.00	26.05	65.65
Magnetite 2	1.85	4.30	32.30	61.55

These two effects will influence the strength of the pellets and even worse pellet reducibility.

Bentonite

The main chemical components and the physical properties of bentonite required for pelletizing experiment were shown in Tables 4 and 5. It can be seen from the Table 4 that Indian bentonite has high iron and low silicon content, which is conducive to iron extraction and silicon reduction. Indian bentonite shows a high Na_2O content (2.92%). In terms of its physical properties, Indian bentonite has the highest methylene blue adsorbed. The montmorillonite content is more than 100%, which is related to the increased adsorption of methylene blue reagent by other clay minerals in bentonite. In addition, the expansion capacity, water absorption index is also very

Table 4 Analysis of chemical composition of bentonite/%

Variety	TFe	SiO_2	Al_2O_3	CaO	MgO	K_2O	Na_2O	S	P	LOI
Indian	11.89	48.27	15.86	1.62	2.69	0.091	2.92	0.059	0.046	10.32

Table 5 Physical properties of bentonite

Variety	Methylene blue adsorbed/g $(100\ g)^{-1}$	Colloid index/%	Swelling capacity/mL g^{-1}	Water absorption (2 h)/%	−0.074 mm/%	−0.043 mm/%	pH
Indian	44.5	99.5	37.0	552.12	98.41	65.32	10.05

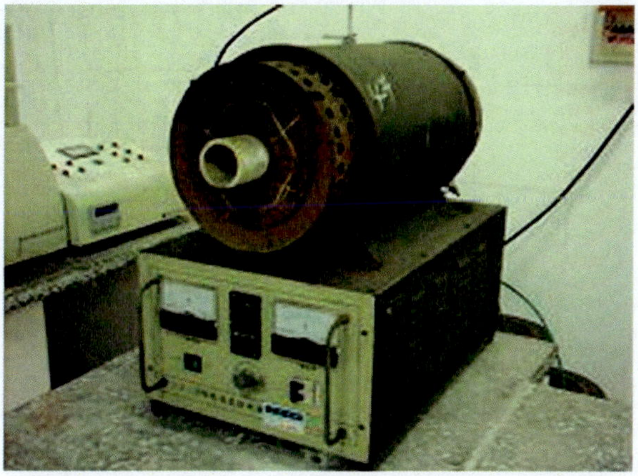

Fig. 1 Horizontal tube furnace. (Color figure online)

high, expansion capacity is 37 mL/g. Bentonite is a binder commonly used in iron ore pellets, which can improve the strength and bursting temperature of pellets but will reduce the iron grade of pellets due to its silicate composition. Therefore, it is usually desirable to reduce the amount of bentonite on the premise of meeting the strength and thermal stability of raw pellets.

Experiment Methods

The small experiment for pellet roasting was carried out in a horizontal tube furnace, including preheating, roasting, and soaking. It is made by linking two tube furnaces, iron chromium aluminum wire resistance furnace for preheating and silicon carbon tube resistance furnace for roasting. During the experiment, the dry balls were installed in a porcelain boat for preheating and roasting experiments (Fig. 1).

Results and Discussion

A small-scale experimental study of pellet tube furnace roasting was carried out.

The research contents include the effect of preheating temperature, preheating time, roasting temperature, roasting time, and grinding fineness on the strength of preheated pellets and roasted pellets.

The qualified raw balls prepared by the experiment were dried in an oven for preheating and roasting. The effect of the preheating system on the quality of the balls with a size of 12 mm was investigated.

Table 6 Effect of preheating temperature on the quality of full MBR preheating balls

Preheating temperature/°C	Preheating time/min	Compressive strength/N P^{1}
950	15	99.6
975	15	146.8
1000	15	147.3
1050	15	212.9
1100	15	270.3
1150	15	418.0

Table 7 Effect of preheating temperature on the quality of 80% MBR-325 mesh 90% + 20% magnetic fine 2 preheating ball

Preheating temperature/°C	Preheating time/min	Compressive strength/N P^{-1}
950	15	338.3
1000	15	477.3
1050	15	568.1
1100	15	766.1
1150	15	1608.6

Effect of Preheating Temperature

The effect of preheating temperature on the strength of preheating balls under different raw material conditions is shown in Tables 6 and 7, respectively. It can be seen that, with increasing preheating temperature, the strength of the preheating ball increases accordingly.

For 100% hematite, with increased preheating temperature, the strength of preheating balls does not increase much. For a mixture of 80% hematite and 20% magnetite, the strength of the preheating ball was highly improved. For the addition of magnetite 2, at the preheating temperature of 1000 °C, the strength of the preheating balls has reached more than 400 N/P. At the preheating temperature of 1150 °C, the compressive strength of the preheating balls exceeds 1000 N/P.

According to the rotary kiln's requirements for the strength of the pellets entering the kiln (more than 400 N/P), it is more appropriate to choose a preheating temperature of 1100–1150 °C. The higher the strength of the pellets entering the kiln, the stronger the ability to withstand mechanical abrasion and erosion,

Effect of Preheating Time

The effect of preheating time on the strength of preheating balls under different raw material conditions is shown in Tables 8 and 9, respectively. It can be seen that the effect of the preheating time on the strength of the preheating ball is much smaller

Table 8 Effect of preheating time on the quality of full MBR preheating ball

Preheating temperature/°C	Preheating time/min	Compressive strength/N P^{-1}
1150	10	394.4
1150	15	418.0
1150	20	453.4
1150	25	437.1

Table 9 The effect of preheating temperature on the quality of 80% MBR-325 mesh 90% + 20% magnetic fine 2 preheating ball

Preheating temperature/°C	Preheating time/min	Compressive strength/N P^{-1}
1150	10	1591.1
1150	15	1608.6
1150	20	1583.6
1150	25	1584.8

than that of the preheating temperature. As the preheating time is extended, the intensity of the preheating ball does not change. Therefore, in order to achieve the necessary strength of the preheating ball, the preheating time should be as short as possible.

Effect of Preheating System on the Strength of Roasted Pellets

Effect of Preheating Temperature on the Strength of Roasted Pellets

Since the preheating strength of all MBR ore powders is low, it is necessary to investigate the effect of preheating temperature on the strength of roasted pellets. The fixed experiment conditions included full pelletized MBR ore powders and the dried raw balls.

The raw ball particle size was 12 mm, the preheating time 15 min, and the roasting temperature/time 1300 °C/15 min, respectively (Table 10).

Table 10 The influence of preheating temperature on the quality of finished ball

Preheating temperature/°C	Preheating time/min	Roasting temperature/°C	Roasting time/min	Compressive strength of preheating ball/N P^{-1}	Compressive strength of finished ball/N P^{-1}
950	15	1300	15	117.8	1190.4
1000	15	1300	15	209.3	1087.9
1150	15	1300	15	369.0	1210.4

Table 11 The influence of preheating temperature on the quality of finished ball

Preheating temperature/°C	Preheating time/min	Roasting temperature/°C	Roasting time/min	Compressive strength of preheating ball/N P^{-1}	Compressive strength of finished ball/N P^{-1}
1150	10	1300	15	371.3	1166.4
1150	15	1300	15	369.0	1210.4
1150	20	1300	15	368.4	1140.0

It can be seen that the preheating temperature has little effect on the compressive strength of the full MBR ore powder roasted pellets. The compressive strength of the finished pellets ranges from 1000 to 1200 N/P. Therefore, for the full MBR ore powder, by increasing of the preheating temperature increase of the strength of the finished pellets could not be reached.

Effect of Preheating Time on the Strength of Roasted Pellets

The size of the raw ball was 12 mm, the preheating temperature 1150 °C, the roasting temperature 1300 °C, and the roasting time 15 min. The effect of the preheating time on the strength of the finished ball is shown in Table 11.

As can be seen from the Table 11 with the extension of the preheating time, the compressive strength, there is only a minor change of the finished pellets, basically maintained at 1100–1200 N/P. For the full MBR ore powder, the preheating time has little effect on the strength of the finished ball. Obviously the roasting system affects the strength of the finished ball.

Effect of Roasting System on the Strength of Finished Pellets

Effect of Roasting Temperature on the Strength of Finished Pellets

It can be seen from the Table 12 that when the roasting temperature is increased, the strength of the finished ball is slightly higher, but the range is very small. Even

Table 12 The influence of roasting temperature on the quality of finished ball

Preheating temperature/°C	Preheating time/min	Roasting temperature/°C	Roasting time/min	Compressive strength/N P^{-1}
1150	15	1270	15	760.5
1150	15	1300	15	1210.4
1150	15	1330	15	1365.6

at higher roasting temperature (e.g., 1330 °C) the strength of the finished MBR ore powder pellets is still 1365.6 N $pellet^{-1}$ only, which is far below the required index of 2500 N/P.

Effect of Roasting Time on the Strength of Finished Pellets

From the Table 13,it can be seen that, with higher roasting temperature of 1330 °C and extended roasting time of 20 min, the strength of the finished MBR ore powder pellets is reduced to a value lower than1000N/P. Therefore, the purpose of increasing the strength of the finished pellets cannot be achieved by extending the roasting time.

From the results above, it can be seen that the full MBR ore is finely ground and pelletized. Under natural alkalinity, the finished ball performance is good, and the preheating roasting performance is poor. The strength of the finished ball is much lower than 2500 N/P. Therefore, without producing flux pellets and adding other substances, the strength of the finished ball can only be increased by ore blending.

Influence of Ore Blending on Quality of Finished Ball

Pelletizing with MBR ore powder with magnetic concentrate 2 ore, raw balls were dried then preheated and roasted. The effects of different proportions of magnetite concentrate on the strength of the finished balls were studied. The results are shown in Table 14. It can be seen that the addition of magnetite concentrate can significantly improve the compressive strength of the finished ball. With the addition

Table 13 The influence of roasting time on the quality of finished ball

Preheating temperature/°C	Preheating time/min	Roasting temperature/°C	Roasting time/min	Compressive strength/N P^{-1}
1150	15	1330	10	734.0
1150	15	1330	15	1365.6
1150	15	1330	20	902.5

Table 14 The influence of proportions of magnetite concentrate on the quality of finished ball

Proportion of magnetite concentration/%	Preheating temperature/°C	Preheating time/min	Roasting temperature/°C	Roasting time/min	Compressive strength/N P^{-1}
0	1150	15	1330	15	1365.6
10	975	15	1300	15	3222.3
20	975	15	1300	15	4948.3
30	975	15	1300	15	4775.5

of magnetite concentrate 2, the addition of 10% can exceed 2500 N/P. Considering the gap between small-scale experiments and expanded experiments, and at the same time and considering the difference in the source of magnetite during on-site production, it is recommended that the ratio of magnetite has to be about 20–30% in the expanded experiments.

Oxidized Pellet Mineral Phase

Mineral Composition of Pellets

Oxidized pellets have high grade, few impurities, and relatively pure mineral composition. The main crystal phase is hematite (Fe_2O_3), followed by a small amount of magnetite (Fe_3O_4), low olivine (CaO, FeO, SiO_2), magnesium ferrite (MgO, Fe_2O_3), aluminum oxide (Al_2O_3), glass, etc. The detailed mineral content is shown in Table 15.

Microstructure of Pellets

Full MBR Ore

Macro view of the ball surface is smooth, black, and gray. The particles are generally 12–14 mm. Under the microscope, the pellets are very dense, the oxidizing atmosphere at the edge of the ball is stronger, and the Fe_2O_3 crystal particles are larger and denser, but they are the primary crystals of the individual particles, with poor crystallization and low strength (see Fig. 2); In the middle and core of the ball, Fe_2O_3 crystals are worse, and the crystal grains are fine (0.03–0.045 mm). However, the olivine particles crystallized completely and clearly (see Fig. 3).

Table 15 Mineral composition of oxidized pellets in Zhuhai (%)

Condition	Hematite	Magnetite	Olivine	Magnesium ferrite	Aluminum oxide	Glass
Full MBR ore	94.61	0.12	0.87	0.15	1.06	3.19
80% MBR + 20% magnetite concentrate 2	92.86	0.75	1.60	0.67	1.08	3.04

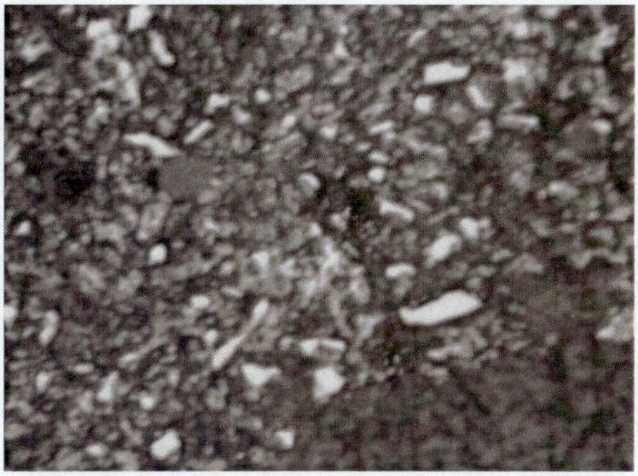

Fig. 2 Crystallization of Fe_2O_3 at the edge of the ball. Fe_2O_3—white, granular; peridotite—gray, granular; holes—black, irregular; resin glue—light yellow-gray, granular; (200 times reflective)

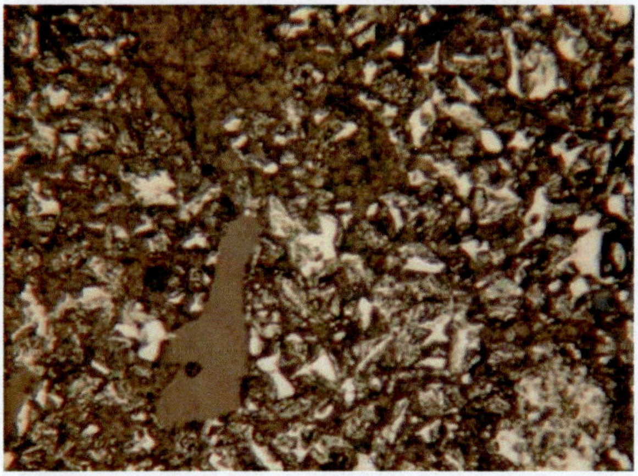

Fig. 3 The distribution of mineral crystals in the core of the ball. Fe_2O_3—white, granular; peridotite—gray, granular; holes—black, irregular; resin glue—light yellow-gray, granular; (200 times reflective). (Color figure online)

80% MBR Ore + 20% Magnetite Concentrate 2

With a certain amount of magnetite in hematite, the microstructure of the pellets is significantly improved. The Fe_2O_3 at the edges of the pellets recrystallizes well, and most of them transition from developed to interconnected crystals. Especially the newly generated Fe_2O_3, which has strong activity and relatively large crystal form, connecting the original Fe_2O_3 together to form the overall structure of the basic interconnection of Fe_2O_3 (shown in Fig. 4). This structure is much stronger than the single granular structure.

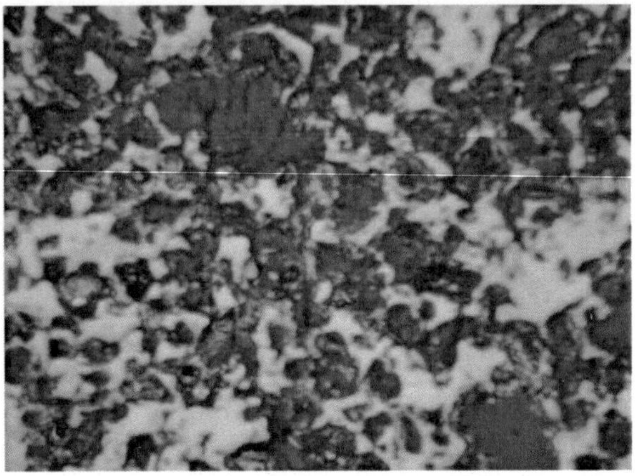

Fig. 4 Crystallization of Fe_2O_3 at the edge of the ball. Fe_2O_3—white, interconnected crystals; peridotite—gray, granular; holes—black, irregular; resin glue—light yellow-gray, granular; (200 times reflective)

Conclusions

(1) The preheating and roasting tests were carried out with MBR ores in the small tube furnace. Under the conditions of natural alkalinity, no matter how optimized the preheating and roasting system is, the compressive strength of the finished ball is only about 1200 N/piece, thus it is too far from the 2500 N/piece required by the index.
(2) The addition of magnetite concentrate can increase the compressive strength of the preheated and finished ball, but the effect of magnetite concentrates from different origins on improving the strength of the finished pellet is quite different. The recommended magnetite ratio is between 20 and 30%.
(3) With a certain amount of magnetite in hematite, the microstructure of the pellets is significantly improved. The Fe_2O_3 at the edges of the pellets recrystallizes well, and most of them transition from developed to interconnected crystals. This positively contributes to a higher strength of the pellets.

References

1. Huang ZC, Zhang YB, Zhu SP et al (2004) Study on mainly made from hematite based oxidized pellets with addition of some magnetite. Iron Steel 4(9–13):59
2. Zhu DQ, Shen WJ, Pan J et al (2011) Research on influence of the alkalinity and the fuel flux to the intensity of hematite pellets. Metal Mine 6:104–108
3. Yin ZG, Tang HQ, Ma XY (2017) Study on intensifying the preparation of oxidized pellets from Iranian hematite concentrates. 2:38–42
4. Fan XH, Yuan XL (2006) Effect of particle size distribution of concentrate on palletizing quality. Chin J Nonferrous Metals 11:1965–1970
5. Zheng HX, Wang Q, Pan XF (2003) Mechanisms of oxidation of magnetite pellet. Sinter Pelletizing 5:13–16

6. Chen YM, Zhang YB (2005) Study on crystallization rule of oxidized pellet. Res Iron Steel 3:10–12
7. Chen YM, Li J (2007) Crystal rule of Fe_2O_3 in oxidized pellet. J Central South Univ 1:70–73
8. Eliasson E, Hooey PL, Annesten H (2007) Formation of potassium slag in olivine fluxed blast furnace pellets. Ironmak Steelmak 5:422
9. Abouzeid A-ZM, Kotb IM, Negm AA (1985) Iron ore fluxed pellets and their physical properties. Powder Technol 3:225
10. Liu WQ (2006) Knowledge and view on development of pellet production in China. Ironmaking 3:10–13
11. Zhou X, Gao X (2010) The analysis on our country pellets industrial development status and trend. Yejin Jingji Yu Guanli 6:16–18
12. Qiao TM, Zheng ZQ, Jiang JZ et al (2001) The plant test of shaft furnace pelletizing with addition of Brazil concentrate in Jinan I&S Co. Sinter Pelletizing 1:15–17
13. Zhao ZW (1999) Review of pellet production for last 30 years at Jinan Iron and Steel Group. Iron Steel 5:1–4
14. Fan XH, Liu C, Chen XL et al (2012) Effects of hematite concentrate ratio on preparing oxidized pellets. Min Metall Eng 5:94–97
15. Gan M, Fan XH, Chen XL et al (2010) Application of Ca, Mg-additives in oxidized pellets. J Central South Univ 5:1645–1651

Metallographic Feature of a Nickel-Based Superalloy in Fluoride Electrolyte Melt

Bowen Li, Xiaodi Huang, and Jiann-Yang Hwang

Abstract Application of a long life and low-cost metallic electrode in salt bath heat treatment, all-electric glass smelters, and primary aluminum production has potential advantages, including great energy saving, significant environmental benefits, adequate electric conductivity, high fracture toughness, excellent thermal shock resistance, and ease of fabrication into complex shapes. Nickel-based alloys have been used in a wide variety of severe operating conditions involving corrosive environment, elevated temperature, high stress, and their combinations. To estimate the feasibility of nickel-based superalloy used as an electrode material, the erosion rate of an Inco nickel-based superalloy in a standard fluoride electrolyte melt was investigated. The diffusion speeds of the metals are in a sequence of Cr > Ni > Fe > Mo. The maximum erosion rate of the superalloy in the fluoride at 1000 °C is approximately 27 μ per hour in a three hours period.

Keywords Nickel-based superalloy · Fluoride · Erosion

Introduction

Immersed or submerged electrode salt bath furnaces are commonly used for heat treatments such as neutral hardening, liquid carburizing, liquid nitriding, austempering, martempering, and other tempering applications. In these furnaces, current is introduced by electrodes and passed through a molten salt to obtain operating temperatures. Currently, the electrodes are made of low carbon steel, heat-resistant alloy, or graphite.

All-electric glass smelters are commonly used to melt granular or powdered raw materials for glass production. Alternating current is introduced by immersed electrodes through a thin layer of glass melt at the furnace bottom to generate heat for

B. Li (✉) · X. Huang · J.-Y. Hwang
Department of Materials Science and Engineering, Michigan Technological University, 1400 Townsend Dr., Houghton, MI 49931, USA
e-mail: boli@mtu.edu

B. Li et al. (eds.), *Materials Engineering—From Ideas to Practice: An EPD Symposium in Honor of Jiann-Yang Hwang*, The Minerals, Metals & Materials Series,
https://doi.org/10.1007/978-3-030-65241-8_10

continuous melting operation. Presently, the electrodes are made of molybdenum, platinum and tin oxide, or graphite.

Primary aluminum is produced using the Hall-Heroult process: passing an electric current between an anode and a cathode in a molten fluoride salt containing dissolved alumina and depositing molten aluminum on the cathode. The current anodes and cathodes are made of graphite to satisfy the requirements for electric conductivity, fluoride erosion resistance, and high temperature strength. The drawback of using a graphite anode is the reaction of the carbon with oxygen discharged from alumina, which releases a great amount of CO_2 into the atmosphere. Molten aluminum is not wetted with a graphite cathode, which prevents narrowing the anode–cathode distance. It is well known that further distance narrowing is capable of reducing energy consumption up to 25% for the primary aluminum production. In the past decade, extensive studies have been carried out to develop inert anode and wetted cathode technologies [1–5]. However, none of the results to date have met the critical requirements.

Several concepts have been proposed to lower the electrochemical reaction temperature from the current 950–750 °C, which provides an opportunity to use metallic anodes and cathodes [5–8]. Successful application of a metallic anode will eliminate the emission of carbon dioxide and perfluorocarbons associated with the current graphite anode, resulting in significant environmental benefits [9, 10]. A metallic anode also has potential advantages of adequate electric conductivity, high fracture toughness, excellent thermal shock resistance, and ease of fabrication into complex shapes for use in an advanced cell design.

Among the above three applications, the electrodes used for primary aluminum production are subjected to the worse environments. The inert anode and wetted cathode material must be electrochemically and thermodynamically stable in the fluoride electrolyte at the operating temperature. Erosion resistance of a metallic anode in the fluoride melt of a primary aluminum cell is the major concern. We chose the "erosion rate" as the criterion for evaluating stability of a potential electrode material in this study.

Nickel-based alloys have been used in a wide variety of severe operating conditions involving corrosive environment, high temperature, high stress, and their combinations. In this study, the erosion rate of an Inco nickel-based superalloy in a standard fluoride melt was investigated using metallographic method.

Materials and Method

The erosion rate experiment was conducted by immersing a nickel-based superalloy in a fluoride melt at an elevated for various times.

The nickel-based superalloy used in this study is 4.75 mm (0.187 inch) thick plate of the Hastelloy and Inco Alloy HX, which is a commercial product purchased from McMaster-Carr Supply Company, Chicago, IL. This superalloy contains 47.5% nickel, 21.8% chromium, 18.5% iron, and 9% molybdenum (ASTM Specification

ASTM B435). It features outstanding oxidation resistance and exceptional strength at temperature up to 1204 °C [11]. It has excellent weldability and is especially suited for use in hot and highly corrosive environments.

The fluoride used in this study is a mixture of aluminum fluoride, sodium fluoride, calcium fluoride, and alumina, provided by Century Aluminum Co. The solid mixture was crushed with a vibrated mill and passed through a 16-mesh screen. The mixture consists of 55.23% aluminum fluoride, 34.42% sodium fluoride, 5.66% calcium fluoride, and 4.69% alumina by weight. According to the ternary phase diagram of AlF_3-CaF_2-Na_3AlF_6, the theoretical melting temperature of this compound is 700 °C [12].

The nickel-based superalloy plate was cut into rectangular specimens of 4.75 mm in width and 50 mm in length. To conduct an erosion test, a specimen was polished to expose fresh surfaces and bedded in 100 g of the fluoride powder contained in a ceramic crucible coated with BN. The BN coating was used to prevent potential reaction between the fluoride and the clay crucible at elevated temperatures. The crucible was then placed in an electric box furnace and heated to 1000 °C. After the furnace reached the desired temperature, the specimen was held at the temperature for a predetermined time. Four holding times at 1000 °C were selected: 0.5, 1.0, 2.0, and 3.0 h. The crucible was then removed out of the furnace and cooled in air to room temperature.

After separated from the fluoride, each superalloy specimen was cut perpendicular to its length and mounted in acrylic resin with the cutting cross-sectional facing bottom. The sample was ground using 180, 240, 400, and 600 mesh sandpapers and finally polished using 6-μ diamond paste and 0.05-μ alumina solution. The polished cross section was then etched with an acid solution.

The etchant was selected according to the ASM Handbook for this type of material and was prepared by mixing equal volume of concentrated nitric acid and glacial acetic acid. The etchant solution was warmed to 60 °C, and the superalloy specimens were swabbed with the etchant for 2 min. The testing procedure flowchart is given in Fig. 1.

The etched specimens were examined under an optical microscope and scanning electron microscope (SEM) equipped with energy dispersive spectrometer (EDS).

Results and Discussion

The crucible containing the fluoride and immersed superalloy specimen after cooling is shown in Fig. 2. From the photo we can see the fluoride bonded on the specimen exposed surface, which indicates good wettability of the fluoride to the alloy during the fluoride molten period. The fluoride solidified after cooling can be broken by impact. The fracture surfaces show lots of inside pores. The superalloy specimen can be readily separated from the broken fluoride mass.

After erosion with fluoride melt, an eroding belt consists of erosion pits was formed and parallelly distributed along with the edges of the alloy. Figure 3 shows a set of optical microstructure pictures of the specimens eroded in the fluoride melt for

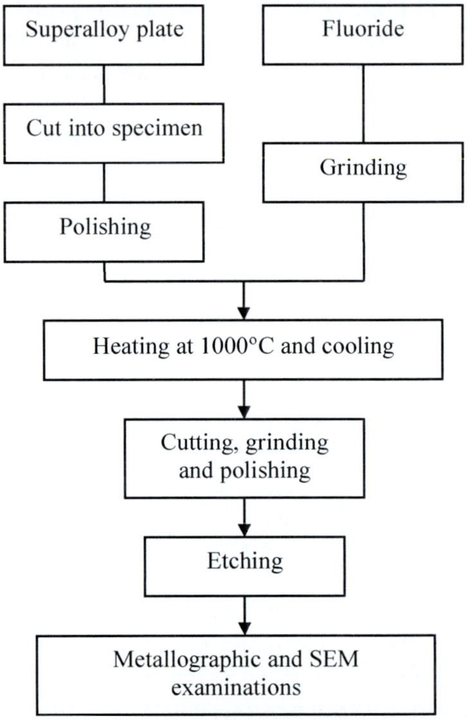

Fig. 1 Test procedure of nickel-based alloy erosion in fluoride melt

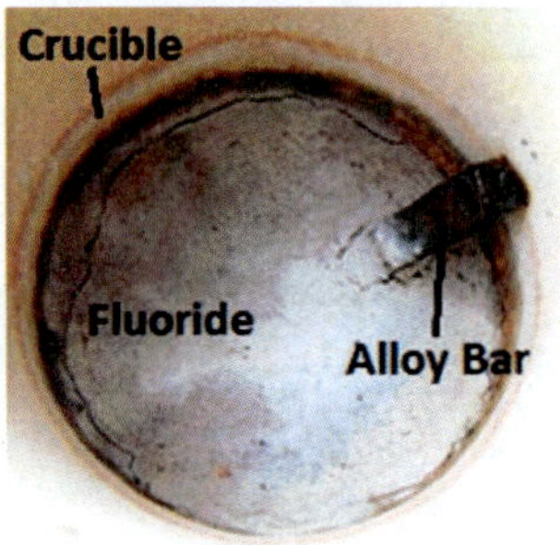

Fig. 2 Ni-based superalloy immersed in solidified fluoride melt. (Color figure online)

different times. As observed under the optical microscope, the microstructure of the specimen with the 0.5-h erosion time shows no obvious difference in the surface and center regions. However, grain boundaries at the specimen surface appear clearly after two hours of erosion. The depth of erosion, represented by the grain boundary exposure, increased with the increase of the erosion time. The depth was less than 10 μ for one hour erosion, but reached 20–50 μ (average 40 μ) for two hours erosion and 55–80 μ (average 67 μ) for three hours erosion.

The erosion rate of the fluoride melt to the superalloy is approximately 27 μ per hour in the three-hour erosion tests (Fig. 4), much higher than an acceptable rate.

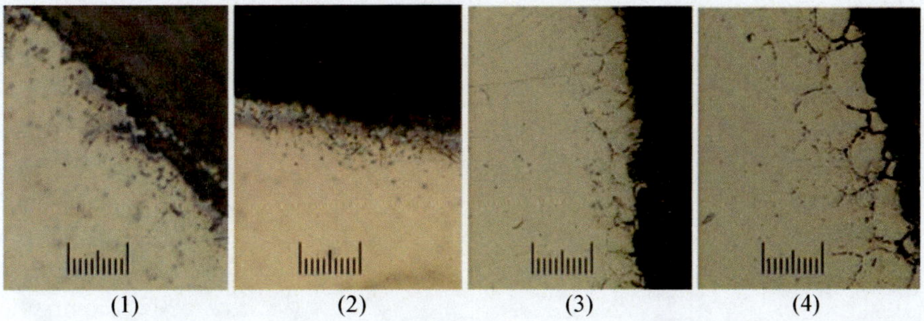

Fig. 3 Erosion of Ni-based alloy in fluoride melts (etched with combined solution of nitric acid and glacial acetic acid) (1) 0.5 h, scale length: 10 μ; (2) 1 h, scale length: 20 μ; (3) 2 h, scale length: 50 μ; (4) 3 h, scale length: 50 μ. (Color figure online)

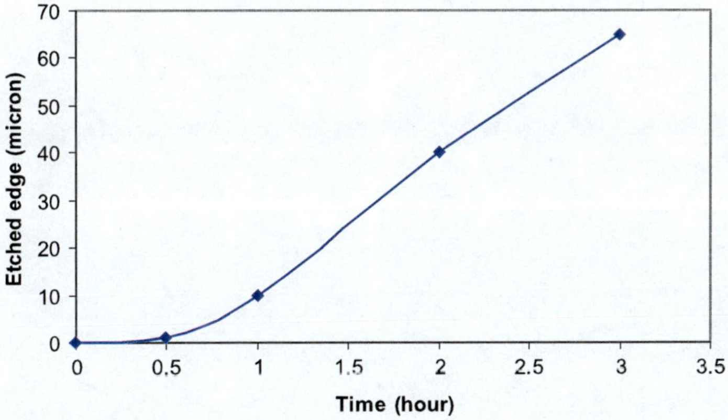

Fig. 4 Erosion rate of Ni-based alloy. (Color figure online)

Figure 5a is the elemental distributions on the cross section of the superalloy specimen after three hours fluoride erosion. The SEM image shows that there is an interface zone between the superalloy and the fluoride (Fig. 5a upper right corner). The EDS (energy dispersive spectrometer) Fe and Cr X-ray dot mapping further revealed existence of the interface zone (Fig. 5b, c). In this zone, the elemental concentrations in an individual grain are homogeneous, but not among the grains which indicates an unstable structure and the metal transitions happened. One can imagine that the erosion started at the most outer surface and also penetrated into grain boundaries adjacent to the surface through atomic solid diffusion in the superalloy and atomic liquid diffusion in the fluoride melt. If the grain boundary erosion is fast in comparison with the surface erosion, one or a few grain layers of solid material could be detached from the specimen body, resulting in unstable interface structure. It might be the case of the superalloy in the fluoride melt. Among the metals, Cr has the fastest diffusion with a critical erosion along grain boundaries which formed

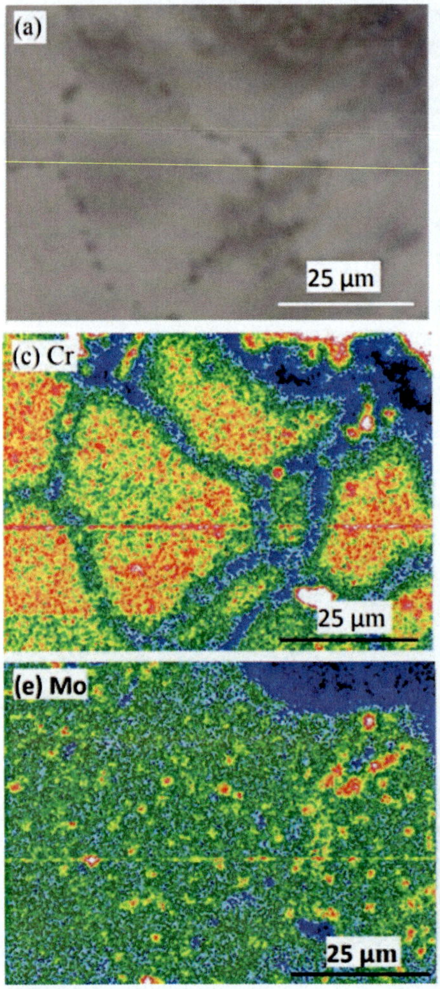

Fig. 5 SEM image and EDS X-ray dot mapping of major elements (eroded in fluoride melt for 3 h) **a** Second electron image; **b** Fe X-ray mapping; **c** Cr X-ray mapping; **d** Ni X-ray mapping; **e** Mo X-ray mapping; **f** F X-ray mapping. (Color figure online)

a wide boundary zone of 2–3 μ in width (Fig. 5c). Mo seems consistent without significant interrupted diffusion. The Mo concentration changes smoothly cross the interfaces of the metal grains (Fig. 5e). Relatively, Ni lost atoms along the grain boundaries, especially at the triple junctions of three grains (Fig. 5d). By comparing the difference of elemental distributions of Fe, Cr, Ni, and Mo, it exhibited that the diffusion speeds of the metals are in following sequence: Cr > Ni > Fe > Mo. In addition, fluorine was deeply diffused into the alloy due to its small atomic size (Fig. 5f).

The superalloy used for this test contains nickel, chromium, iron, and molybdenum. Nickel is a good erosion resistance material. It offers excellent erosion resistance to non-oxidizing acids and caustic alkalis. In a nickel-based alloy, chromium imparts improved resistance to oxidizing media such as nitric acid (HNO_3) and hot H_3PO_4. Molybdenum in nickel substantially improves resistance to non-oxidizing acids. However, iron is typically used to reduce costs, not to promote erosion resistance [13, 14]. Therefore, at high temperature and contacting with fluoride melt, the chromium and iron atoms on the surface and grain boundaries would be easily attacked and diffused into the molten fluoride. Based on this experiment, the chosen superalloy did not show promising erosion resistance. A nicked-based alloy without iron content could be further investigated. In addition, grain size effect and correlation between erosion depth rate of grain boundary and apparent erosion rate of specimen need to be investigated as well.

Summary

Nickel-based superalloy offers excellent corrosion resistance at high temperature against air and general compounds. However, the superalloy used in this study did not show great potential to be used as an electrode material contacted directly with fluoride melt. The diffusion speeds of the metals are in a sequence of Cr > Ni > Fe > Mo. The maximum erosion rate of the fluoride melt to the superalloy is approximately 27 μ per hour in the three-hour erosion tests at 1000 °C. This erosion rate cannot meet the requirement for the current Hall-Heroult process.

Acknowledgements This research was sponsored by the U.S. Department of Energy (DOE) Award through a contract with Michigan Technological University (DE-FC36-02ID14402).

References

1. Brown GW, Beck TR, Frizzle PB (2002) Cathode connector for aluminum low temperature smelting cell. US Patent 6,419,813
2. Bradford DR, Barnett RJ, Mezner MB (2003) Anode for use in aluminum producing electrolytic cell. US Patent 6,558,525
3. Liu X, Ray SP, LaCamera AF et al (2003) Synthesis of multi-element oxides useful for inert anode application. US Patent 6,419,813
4. Rapp RA (2003) Development of a novel non-consumable anode for electrowinning primary aluminum. Report DOE-OIT Project DE-FC07–99ID13813, US
5. Rioult F, Pijolat M, Valdivieso F (2006) High-temperature oxidation of a Cu–Ni based cermet: kinetic and microstructural study. J Am Ceram Soc 89:996–1005
6. Brown CW (2003) In-situ observations of frozen electrolyte in laboratory reduction cells. Light Metals 2003, The Minerals, Metals & Materials Society, Warrendale, PA, US, pp 293–297
7. LaCamera AF (1989) Electrolysis of alumina in a molten salt at 760 °C. Light Metals 1989, The Minerals, Metals & Materials Society, Warrendale, PA, US, pp 291–295

8. Beck TR (1994) Production of aluminum with low temperature fluoride melts. Light Metals 1994, The Minerals, Metals & Materials Society, Warrendale, PA, US, pp 417–423
9. The Aluminum Association, Inc (1998) Inern Anode Roadmap. The Aluminum Association, Inc., Washington, DC, US
10. Pawlek RP (2002) Inert anodes: an update. Light Metals 2002, The Minerals, Metals & Materials Society, Warrendale, PA, US
11. McMaster-Carr Supply Company, Nickel high-temperature alloy (production manu), www.mcmaster.com. Accessed Sept 2015
12. Craig DF, Brown JJ (1980) Phase equilibria in the system CaF_2-AlF_3-Na_3AlF_6 and part of the system CaF_2-AlF_3-Na_3AlF_6.Al_2O_3. J Am Ceramic Soc 63:254–261
13. Klarstrom DL (1987) Characteristics of nickel-bas alloys, vol 13. ASM Handbook, Erosion, ASM International, Materials Park, OH, US, pp 641–643
14. Sridhar N (1987) Behavior of nickel-base alloys in corrosive environments, vol 13. ASM Handbook, Erosion, ASM International, Materials Park, OH, US, pp 643–647

Concurrent Production of Iron and Syngas from Iron Ore and Coal Mixture

Xiaodi Huang, Jiann-Yang Hwang, and Rick Kauppila

Abstract This study is to explore the feasibility of iron and syngas co-production from iron ore and pulverized coal mixture in an enclosed vessel using microwave heating. Reactions take place after the materials are subjected to heating above certain temperatures. Coal serves as the reducing agent for iron oxide and the source of carbon and hydrogen for the syngas generation. The iron oxide serves as the iron source for steelmaking as well as the oxygen source for the carbon partial reaction to form CO. In comparison with conventional coal gasification, the oxygen comes from iron oxide instead of pure oxygen separated from the air. This study showed that direct reduced iron with over 95% metallization and syngas with total concentration of CO and combustibles over 80% were readily produced.

Keywords Microwave · Steelmaking · Iron · Syngas

Introduction

Steel Production

Steel production is one of the most energy intensive processes, accounting for 7% of total industrial energy consumption in the United States [1, 2]. The current steelmaking is not environmental friendly because of necessary coking involved. Great energy saving and emission reductions in steel production are desired. Currently, steel is produced by two types of operations: integrated mills and minimills. Integrated mills utilize combinations of blast furnaces (BF) and basic oxygen furnaces (BOF) to produce steel from iron ore. Steel produced from this route is of higher

X. Huang (✉) · J.-Y. Hwang
Department of Materials Science and Engineering, Michigan Technological University, Houghton, MI, USA
e-mail: 18201560720@139.com

R. Kauppila
U.P. Steel, Houghton, MI, USA

B. Li et al. (eds.), *Materials Engineering—From Ideas to Practice: An EPD Symposium in Honor of Jiann-Yang Hwang*, The Minerals, Metals & Materials Series,
https://doi.org/10.1007/978-3-030-65241-8_11

quality. Minimills employ electric arc furnaces (EAF) to remelt steel scrap with direct reduced iron (DRI) for impurity dilution. Steel produced from a minimill is generally lower in quality. Integrated mills and minimills each take about 50% of the steel market. In integrated mill operations, sintered iron ore pellets, coke, and lime are charged into a blast furnace. Heated air is blown in at high velocity (20 m/s) to partially combust the coke to generate carbon monoxide and heat [3–8]. The iron ore pellets are reduced by the carbon monoxide and melted to form liquid iron. The exhaust gas containing up to 40% CO and 6% H_2, or typically 25% CO and 13% CO_2 is released into the environment at a temperature of about 500 °C [9, 10]. The liquid iron is then sent to a BOF where pure oxygen is blown into the iron at a supersonic velocity [5] to remove excessive carbon and convert liquid iron into liquid steel. This production route is associated with many environmental issues because of the link to (1) inherent demand for coke production (one of the most polluting operations), (2) discharge of great amounts of particulate, CO_2, CO, SO_2, NO_x, VOCs, and slag, and (3) high energy consumption. The steel industry has invested over $7 billion over the last 30 years for environmental control, which represents approximately 15% of the total operating cost of the industry [10].

Syngas Production

Syngas is a mixture of carbon monoxide and hydrogen generated by the gasification of a carbon containing material. It can be used as the intermediate to produce synthetic gasoline and diesel. The original process was invented by two German researchers and named the Fischer-Tropsch process. This process was used by Germany during World War II to produce synthetic fuel from coal. Sasol, a South Africa-based company, has also used this technology to produce synthetic diesel from coal since the 1950s due to the world isolation of South Africa. This process has attracted special interests in coal-rich but petroleum-poor countries [11, 12]. A relatively newer development called Integrated Gasification Combined Cycle (IGCC) [12, 13], aiming at increasing power plant energy efficiency and reducing greenhouse emissions, utilizes the same coal gasification technique to produce syngas from coal. The syngas is then used to drive both a gas turbine and a steam turbine to generate electricity. Coal gasification is accomplished by using a gasifier which is similar to a combustor. The difference is that a gasifier uses oxygen instead of air for partial combustion. The amount of oxygen inside the gasifier is carefully controlled so that only a portion of the carbon containing material burns. It is desired to lower the gasification equipment cost, simplify the operation, and make it suitable for various coals [14, 15].

Alternative Approach

Our intention is to develop a new technology which can save steelmaking energy and reduce environmental impact, while at the same time producing syngas as the second primary product. This technology utilizes the advantages of rapid volumetric heating and higher energy efficiency through the use of microwave heating. The concept lies in the fact that iron ore and carbon are excellent microwave absorbers [16, 17]. In this new process, iron ore is crushed, ground, and concentrated by conventional processes. The concentrated iron oxide is mixed with pulverized coal and lime as the feedstock. The coal serves as the reducing agent for iron oxide, an auxiliary heating source via an exothermal oxidation reaction, and the carbon and hydrogen source for the syngas production. The iron oxide serves as the source of iron units for steel production as well as the source of oxygen for the carbon partial reaction to form CO. In comparison with conventional coal gasification, the oxygen comes from iron oxide instead of from pure oxygen produced by an oxygen plant. In the new technology, the entire heating and reaction takes place in an enclosed system. Because no air is required for combustion, the coal volatiles (primarily methane, CH_4) and CO comprise the majority of the exhaust gas and can be readily controlled. Lime is used as a fluxing agent as well as a desulphurization agent for the steel production.

Experimental

Apparatus

Figure 1 illustrates the device we built and used for the study. The device consists of a 1 kW microwave oven, a fireclay crucible, a steel cover welded with a steel stem, a thermometer, a cotton filter, three flow meters, and three gas monitors. This device was used to heat the materials in the crucible by microwave radiation and measure the off-gas temperatures, sample final temperatures, flow rates of off-gases, and concentrations of CO, combustibles, H_2S, and CO_2 in the off-gases.

Gas Monitoring

A GasAlertMicro monitor manufactured by BW Technologies, a PCA2 analyzer manufactured by Bacharach, and a CO_2 monitor manufactured by RKI Instrument, Inc. were used. The combination of the three gas monitors can detect CO, combustibles (hydrogen, methane, propane, n-butane, n-pentane, n-hexane, methanol, ethanol, isopropyl alcohol, acetone, methyl ethyl ketone, and toluene),

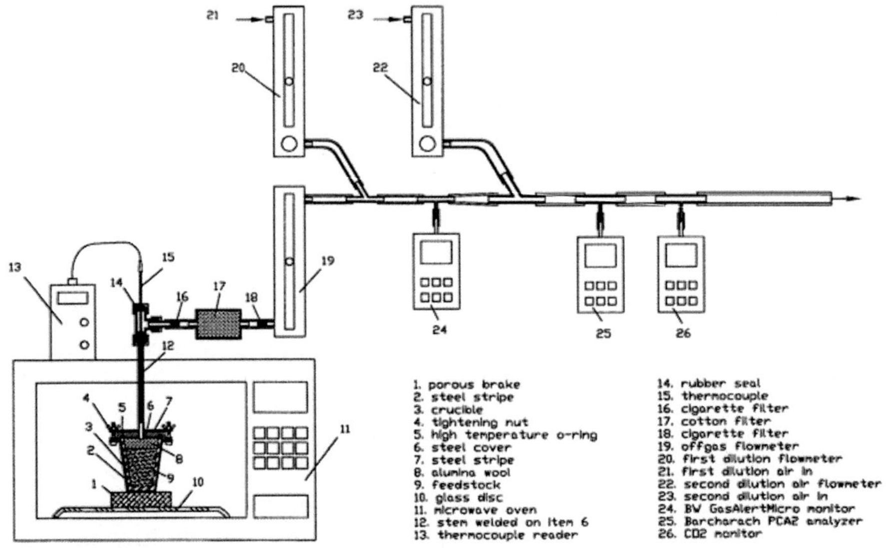

Fig. 1 Apparatus for co-generation of DRI and syngas

H_2S, and CO_2. After received, the three gas monitors were modified in order to use plug-in electricity instead of using batteries for continuous monitoring.

Off-gas Temperature and Sample Temperature

To measure the off-gas temperature, a thermocouple was inserted into the stem and kept its tip one inch from the stem opening as illustrated in Fig. 1. The one inch distance can avoid microwave irradiation into the stem and influence the temperature reading. After microwave heating, the thermocouple was slide down into the feedstock to measure the sample final temperature.

Off-gas Flow Rate

Cole-Parmer N-082-03 flow meter and Matheson E801 flow meter were used to measure the flow rate of off-gas during microwave heating depending on the amount of off-gas generated. The former was used for lower flow rate and the later for higher flow rate.

Table 1 Compositions of ores, %

Ore	Fe	SiO_2	CaO	MgO	MnO	Al_2O_3	P	K_2O
Magnetite	62.9	4.92	4.81	1.60	0.12	0.12	0.015	0.016
Hematite	61.6	4.51	4.31	1.62			0.034	

Calibrations

The three gas monitors were calibrated several times during their services to the project using standard calibration gases purchased from the monitor suppliers. The flow meters were verified with their flow rate charts at beginning and also calibrated several times during their services to the project using an Aalborg GFM17 electronic flowmeter and comparison with their initial conditions.

Dilution

The detection upper limits of CO and combustibles sensors are much lower than the concentrations of CO and combustibles in the produced syngas. Dilution has to be introduced in order to use the gas monitors. Due to the sensor types, inert gases cannot be used, instead, we used compressed air as the dilution gas. Different sensors have different detection limits. Two steps of dilution were applied to adjust the dilution ratios for different monitors.

Raw Materials

A typical magnetite concentrate and a typical hematite concentrate produced by CCI (Cleveland Cliffs Inc.) were used as our iron ore testing materials. Table 1 lists their compositions. A western coal and an eastern coal (warehouse coal identified herewith) were used as the source of reducing agent and combustibles. Table 2 lists their analysis results which were made by Commercial Testing & Engineering Co. For comparison purpose, a synthetic graphite powder with low sulfur content and no combustibles was also used as the reducing agent and the material for syngas generation. The graphite powder was supplied by Sigma-Aldrich with ID 282863.

Chemical Analyses

The total iron content of a DRI sample was determined by a regular digestion method using an ICP (Inductively Coupled Plasma, Leeman Labs Incorporation, and Model

Table 2 Composition of coals, %

	Western coal		Warehouse coal	
	As received	Dry basis	As received	Dry basis
Proximate analysis				
Moisture	1.61	–	12.66	–
Ash	9.73	9.89	8.05	9.22
Volatile	29.76	30.25	38.72	44.33
Fixed carbon	58.90	59.86	40.57	46.45
Btu/lb	13,230	13,446	9041	10,352
Sulfur	1.01	1.03	0.79	0.91
MAF Btu		14,922		11,403
Ultimate analysis				
Moisture	1.40	–	12.66	–
Carbon	72.72	73.75	55.15	63.14
Hydrogen	4.76	4.81	3.63	4.16
Nitrogen	1.38	1.40	0.79	0.90
Sulfur	1.31	1.33	0.79	0.91
Ash	10.68	10.83	8.05	9.22
Oxygen	7.77	7.78	18.93	21.67

PS3000) at Michigan Tech. Metallization of DRI was determined using a method accepted by major steel companies. This method is described below. After cooled down to the room temperature, a DRI sample was weighed, ground repeatedly, and sieved through No. 100 mesh copper sieve. If some large ductile iron particles were found, they were taken out and weighed. The remaining DRI was then used to prepare a solution. Ten grams of the powdered DRI were taken and dumped into a beaker, followed by adding ten grams of $CuSO_4$ reagent pure powder and 300 ml distilled water. After stirred and heated at 100 °C for about 30 min, the slurry was filtered into a 1000 ml flask. Distilled water was added to dilute the solution to 1000 ml. The diluted solution was then ready for the analysis of metallic iron concentration using an ICP. ICP requires individual element concentration in the solution less than 100 ppm, therefore the solution may be further diluted before analysis. During the process of solution preparation, the following chemical reaction took place and all reduced iron in the sample was dissolved through substituting copper ions.

$$Fe + Cu^{2+} = Fe^{2+} + Cu$$

The metallization of the sample was derived from the total iron content of the sample, the metallic iron concentration in the diluted solution, the total weight of the sample, and the weight of large iron particles if found.

The carbon and sulfur contents of the DRI samples were determined using Leco C&S analyzer CS-244 at Michigan Tech.

Test Procedure

Selected coals were pulverized and grinded until all passed 100 mesh. The pulverized coals and iron ore concentrates were dried in an oven at 105 °C for over 24 h and sealed in plastic bags after cooling for later uses. To prepare a sample, 100 parts of iron ore and predetermined ratios of coal or graphite powder were mixed up and down in a plastic bottle with a couple of ½″ ceramic balls for 5 min. Table 3 gives the test/sample IDs and the mixing recipes. Coal #1 is our warehouse coal and Coal

Table 3 Sample ID and mixing recipes

Test/Sample ID	Iron ore	Coal/Carbon	Coal/Ore weight ratio
M-Coal1-15	Magnetite	Warehouse	0.15
M-Coal1-20	Magnetite	Warehouse	0.2
M-Coal1-25	Magnetite	Warehouse	0.25
M-Coal1-30	Magnetite	Warehouse	0.3
M-Coal2-15	Magnetite	Western	0.15
M-Coal2-20	Magnetite	Western	0.2
M-Coal2-25	Magnetite	Western	0.25
M-Coal2-30	Magnetite	Western	0.3
M-Graphite-15	Magnetite	Graphite	0.15
M-Graphite-20	Magnetite	Graphite	0.2
M-Graphite-25	Magnetite	Graphite	0.25
M-Graphite-30	Magnetite	Graphite	0.3
H-Coal1-15	Hematite	Warehouse	0.15
H-Coal1-20	Hematite	Warehouse	0.2
H-Coal1-25	Hematite	Warehouse	0.25
H-Coal1-30	Hematite	Warehouse	0.3
H-Coal2-15	Hematite	Western	0.15
H-Coal2-20	Hematite	Western	0.2
H-Coal2-25	Hematite	Western	0.25
H-Coal2-30	Hematite	Western	0.3
H-Graphite-15	Hematite	Graphite	0.15
H-Graphite-20	Hematite	Graphite	0.2
H-Graphite-25	Hematite	Graphite	0.25
H-Graphite-30	Hematite	Graphite	0.3

#2 is a western coal. After separation from the ceramic balls, the mixed material was charged into a fireclay crucible and enclosed with a cover. The fireclay crucible top edge was grinded before use on a metallographic rotating disc to make the top edge surface flat and smooth for air tight seal with a high temperature O-ring, which was embedded in the steel cover groove. After each heating test, the crucible was carefully inspected for cracks and replaced with a new one if a crack was found. Each sample in the crucible was heated by microwave for 10 min. Two persons operated each test. Off-gas temperature, flow rate, and composition readings were recorded every 30 s. If the readings were unusual, the entire system was checked carefully for potential problems. After correction, the test was repeated. There were several air filters in the line of the off-gas to the flow meter and gas monitor probes. The air filters were replaced if their white color changed obviously.

Results and Discussion

Combustibles Generation

Concentration of combustibles in the produced syngas (off-gas) varies with different iron ores, coals, and mixing ratios while heating. From the samples with a coal we can see that the concentrations of combustibles increase quickly and reach the highest in the first 3–4 min and then decrease gradually. The highest concentration is around 50% for the magnetite samples and 35% for the hematite samples. We believe the difference is due to higher heating rate of magnetite samples since magnetite is a better microwave absorbing material than hematite. The samples containing warehouse coal show obvious higher concentrations of combustibles near the end of heating in comparison with the samples containing western coal. It can be explained by higher volatiles in warehouse coal than in western coal. It is naturally thought that the samples with higher ratio of coal should have higher concentrations of combustibles. Several curves with the lowest ratio 0.15 and the highest ratio 0.30 do show this trend. However, this trend is not always consistent. Higher coal ratios should increase output of combustibles but may increase CO output as well. As the result, the concentration of combustibles may not increase. The hematite samples with graphite show zero combustibles, but the magnetite samples with graphite show significant combustibles, which is a surprise. A possible reason is that the magnetite ore concentrate has some remaining organic reagent added during the mineral process.

CO Generation

CO concentration of the produced syngas varies using different iron ores, coals, and mixing ratios while heating. CO shows up almost immediately after microwave

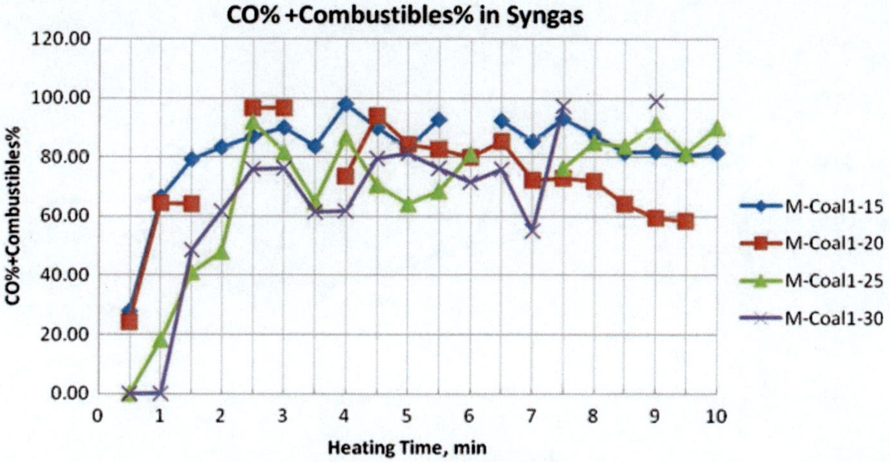

Fig. 2 CO + combustibles of magnetite + coal1 samples. (Color figure online)

heating the magnetite samples. There is some delay of CO appearance in the syngas of hematite samples. We believe this difference is due to the fact that magnetite is a better microwave absorbing material. The temperatures were higher in magnetite samples during the early microwave heating. Later on magnetite and hematite changed to Wustite and then DRI during reduction. The difference should be gradually reduced. In general, we can see very high CO concentrations in the syngas, 30–70%, mostly over 50% near the end of heating. Samples with graphite are more consistent with high CO concentrations.

CO + Combustibles

Figures 2, 3, 4, 5, 6, 7, 8, 9 and 10 show the total CO + combustibles, the energy value. The concentration of total CO + combustibles can be very high over 80% in the produced syngas if enough carbonaceous material was blended into the feedstock. Different carbonaceous materials have different minimum amount needed for good quality of syngas generation. For comparison purpose, Tables 4 give a few examples of syngas produced by the current coal gasification technologies. We can see that our syngas has the potential to surpass those gases in quality due to lower CO_2.

DRI Metallization

Figure 11 shows the metallization analysis results of several samples. The metallization increases as addition of a carbonaceous material increases. The samples with

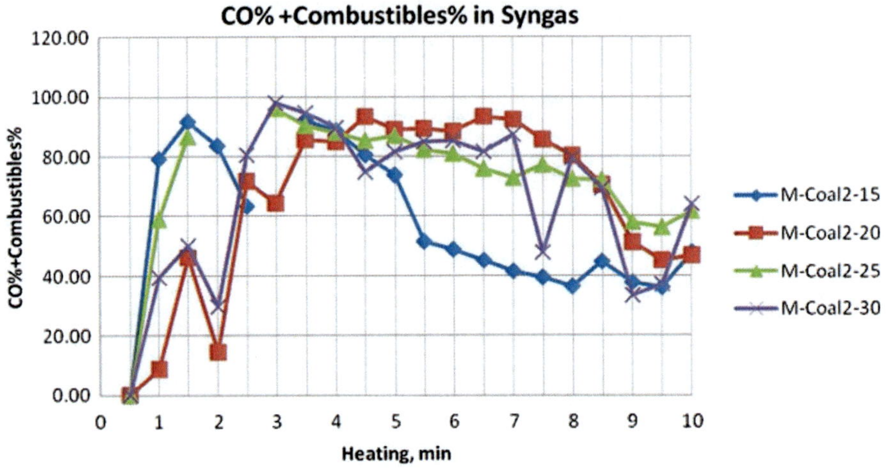

Fig. 3 CO + combustibles of magnetite + coal2 samples. (Color figure online)

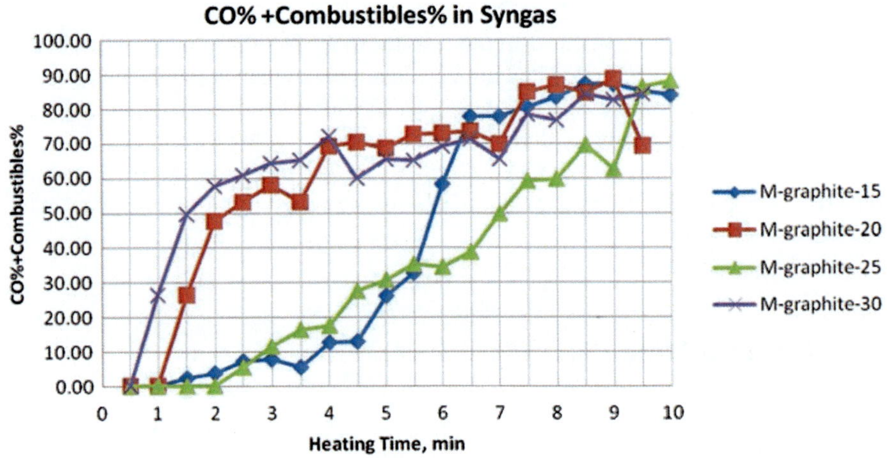

Fig. 4 CO + combustibles of magnetite + graphite samples. (Color figure online)

graphite reach 80–100% metallization when the graphite to ore ratio is over 0.2 and then the metallization remains flat as the graphite increases further. The samples with coals reach about 70% metallization when the coal to ore ratio is 0.3 and the curves do not show flat trend yet. More coal should be added to increase the metallization. It is obvious that less addition of graphite is needed than coal to achieve the same level of metallization. It has been proved by our previous research that the fixed carbon is the real reducing agent. The graphite has basically 100% fixed carbon and the warehouse coal only 46.45% fixed carbon and the western coal 60% fixed carbon.

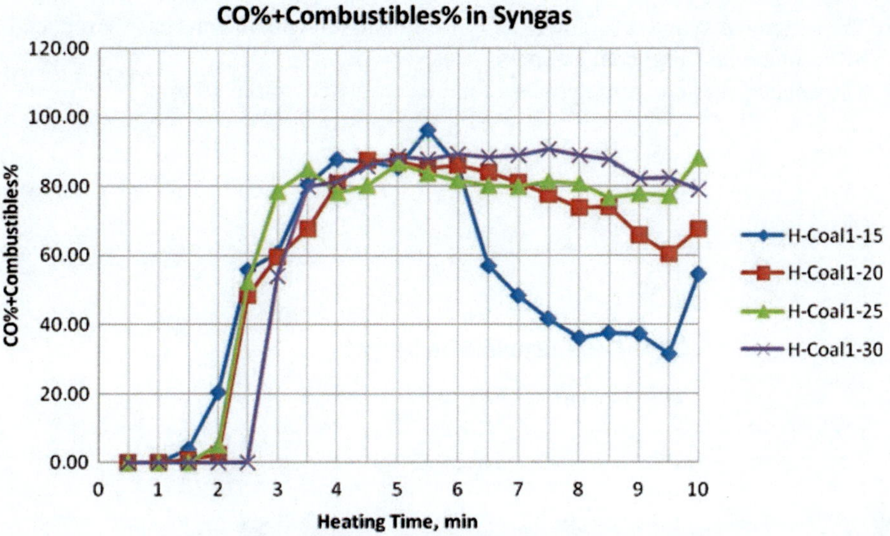

Fig. 5 CO + combustibles of hematite + coal1 samples. (Color figure online)

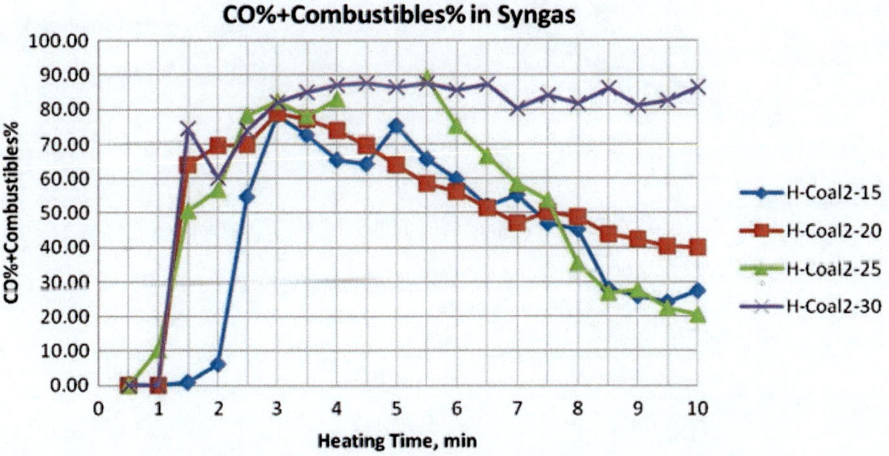

Fig. 6 CO + combustibles of hematite + coal2 samples. (Color figure online)

Conclusions

- DRI and syngas can be produced from coal and iron ore mixtures in an enclosed vessel heated by microwave.
- Metallization of the produced DRI can be over 95%.
- The total concentration of CO and combustibles in the produced syngas can be over 80%.

- The generated syngas is of lower CO_2 in comparison with the raw gases produced by commercial Lurgi coal gasifiers.
- The concurrent generation process can be completed within 10 min.

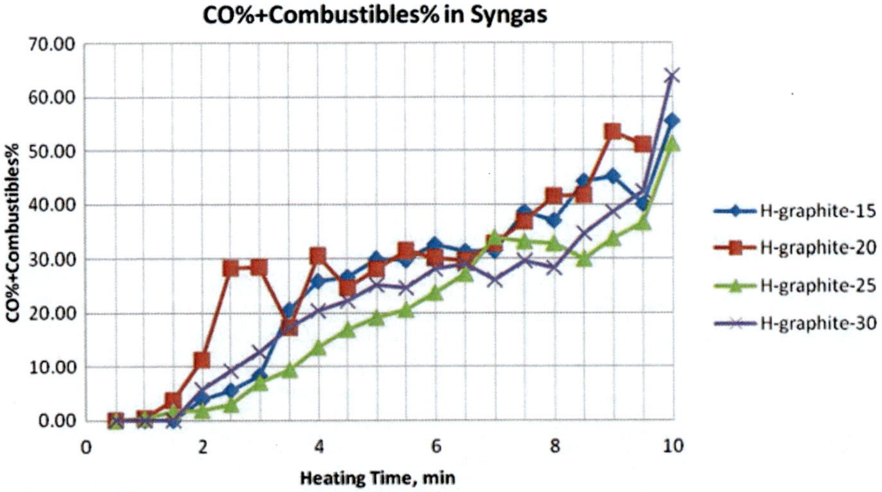

Fig. 7 CO + combustibles of hematite + graphite samples. (Color figure online)

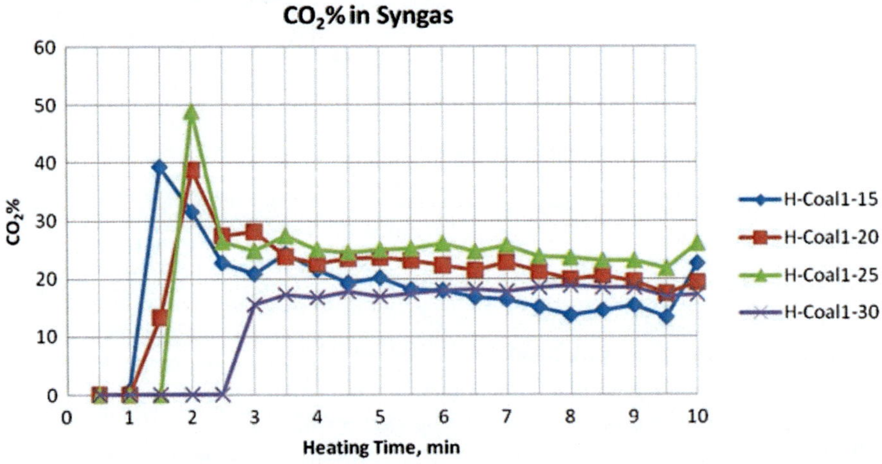

Fig. 8 CO_2 of hematite + coal1 samples. (Color figure online)

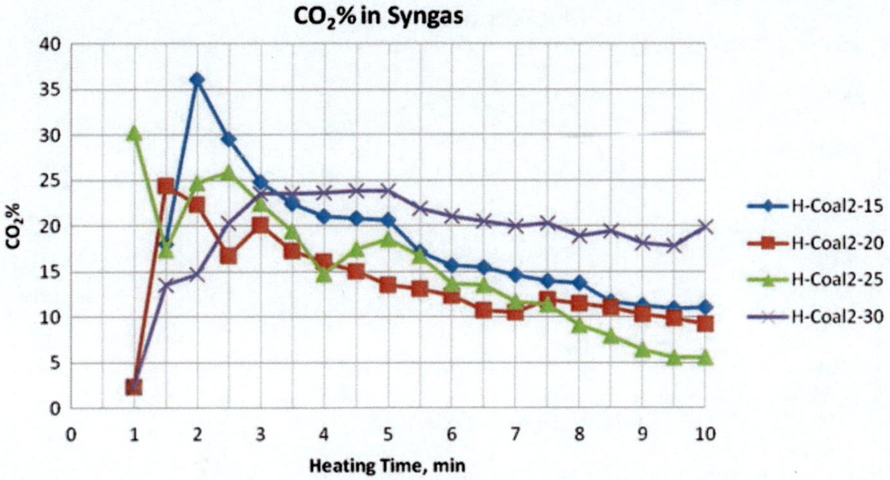

Fig. 9 CO_2 of hematite + coal2 samples. (Color figure online)

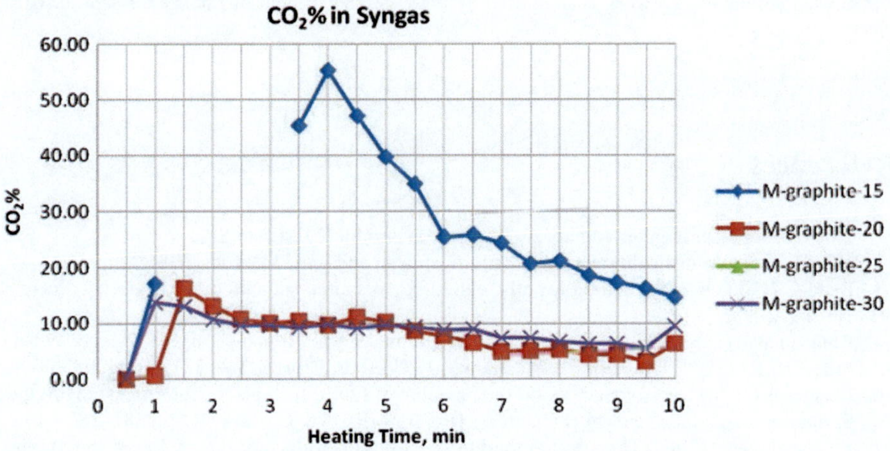

Fig. 10 CO_2 of magnetite + coal2 samples. (Color figure online)

Table 4 Average of raw gas analysis by 14 Lurgi Mark IV Gasifiers*

Component	Concentration (%)
H_2	39
CO_2	32
CO	15
Methane	12
C_2+	0.8
H_2S	0.7
BTU/scf (HHV)	315

*Dakota Gasification Company

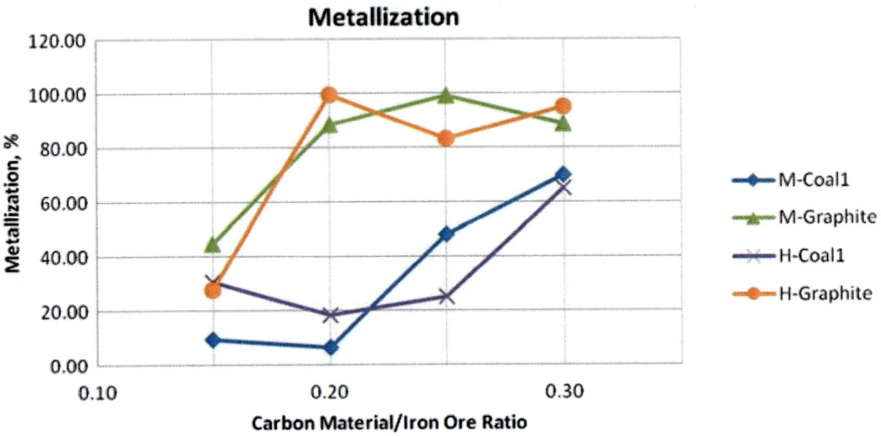

Fig. 11 DRI metallization. (Color figure online)

Acknowledgements This work is supported by the U.S. Department of Energy through a SBIR Phase I project.

References

1. (2001) Steel technology roadmap. American Iron and Steel Institute
2. (1999) Annual statistical report 1998. American Iron and Steel Institute (AISI)
3. Bisio G (1993) Energy method for efficient energy resource use in the steel industry. Energy 18(9):971–985
4. Bouman RW (1983) Energy requirements for ironmaking and steelmaking. I&SM, pp 23–25
5. Fruedhan RJ (1999) Evaluation of steelmaking processes. DOE/ID12847-5 (DE94005368)
6. Eketorp S (1978) Decisive factors for the planning of future steel plants. In: Proceeding of the third international iron and steel congress, 16–20 April 1978, Chicago, IL, pp 181–185
7. Fruehan RJ et al (2000) Theoretical minimum energies to produce steel for selected conditions. Report for the U.S., Department of Energy, Office of Industrial Technologies
8. Stubbles J (2000) Energy use in the U.S. Steel Industry: a historical perspective and future opportunities. Report for the U.S., Department of Energy, Office of Industrial Technologies
9. Kakela P (1979) Pelletized vs. natural iron ore technology: energy, labor, and capital changes. Energy Conserv 18–19
10. Margolis N, Brindle R (2000) Energy and environmental profile of the U.S. Iron and Steel Industry. Report for the U.S., Department of Energy, Office of Industrial Technologies
11. Coal to clean fuel—the Shenhua investment in direct coal liquefaction. In: Twenty-first annual international Pittsburgh coal conference, 13 September 2006
12. (2005) Coal gasification 101. Electric Power Research Institute
13. Syngas refiner. Market and technology analysis for syngas-derived products from any hydrocarbon source, vol III, No 1, 1 January 2007
14. Texaco gasification process. EPA 540/R-94/514a, April 1995, 17, 2004. Grand Cube Osaka (Osaka International Convention Center), Osaka, Japan
15. Shell coal gasification process for power and hydrogen/chemicals. GTC October 2004 Washington

16. Brown DC (1963) Direct reduction of iron ores—a 1961 review. Recent advances in ferrous metallurgy, Symposium Series 43, vol 59. American Institute of Chemical Engineers, pp 53–83
17. Chabinsky IJ (1988) Applications of microwave energy past, present and future. Mat Res Soc Symp Proc 124:17

The Formation Mechanism of the Third Phase in Nickel Electrolyte

Chen Ailiang, Mao Jiale, Luo Guanwen, Lu Sujun, Zhang Peng, Ma Yutian, Chen Shengli, Du Zuojuan, Bowen Li, and Qiao Jinxi

Abstract The third phase is generated inevitably, but affects the quality of the electric nickel product in the nickel metallurgical process. This paper studies the formation and the prevention mechanism of the third phase in the raffinate. The analysis of Fe-S–Si-H_2O system shows that iron exists in the form of Fe^{3+} at the pH value ≤ 2.17. With pH increasing, Fe^{3+} begins to hydrolyze to form $FeOH^{2+}$. $FeOH^{2+}$ continues to hydrolyze to form $Fe(OH)_2{}^+$ at the pH value ≥ 4.72. Fe^{3+} is reduced and hydrolyzed to $FeOH^+$ when the potential ≤ 0.21 V. The precipitation of Fe_2SiO_4 is reduced at the potential ≤ 0.35 V and $5 \leq$ pH value ≤ 13.2 when the solution contains silicon dioxide. The analysis of the third phase shows that it is mainly iron hydroxide, iron polynuclear hydroxyl complex ions, and Fe-SO_4 complexes. It provides theoretical guidance for the formation of the third phase in the extraction multiphase system. This also plays a positive role in the production of high-quality nickel products.

Keywords Nickel electrolyte · Third phase · Formation mechanism

Introduction

Extraction separation technology plays an important role in the process of metal purification. Many high purity metals can be obtained by the process of leaching-extraction-electrodeposition. Nickel electrodeposition is produced with this method [1]. At present, the extraction method is widely used in the electrolyte purification process. This leads to the residue of organic compounds such as extractant oil in the

C. Ailiang · M. Jiale · L. Guanwen · D. Zuojuan · Q. Jinxi (✉)
School of Metallurgy and Environment, Central South University, Changsha 410083, China
e-mail: Qiaojinxi@csu.edu.cn

L. Sujun · Z. Peng · M. Yutian · C. Shengli
State Key Laboratory of Nickel and Cobalt Resources Comprehensive Utilization, Jinchang 737100, China

B. Li
Michigan Technological University, Houghton M49931, USA

B. Li et al. (eds.), *Materials Engineering—From Ideas to Practice: An EPD Symposium in Honor of Jiann-Yang Hwang*, The Minerals, Metals & Materials Series,
https://doi.org/10.1007/978-3-030-65241-8_12

solution. On the one hand, the continuous accumulation of oil will lead to high carbon content and blacken the surface of electro nickel in the process of electrodeposition. On the other hand, the current efficiency of the electrodeposition process was reduced by the oil. Moreover, the oil increases the precipitation potential of nickel and lead to hydrogen precipitation, resulting in pores on the surface of the nickel plate [2].

The extractants like Cyanex272, P204, and P507 are often used to remove impurities from nickel solution during the process of leaching-extraction-electrodeposition. This kind of extraction organic phase and inorganic phase are immiscible. It is very easy to form oil-in-water and water-in-oil emulsion with small droplets in another phase. Moreover, a small part of silicon dioxide in the ore will also enter the solution in the process of extracting nickel from nickel ore. This increases the solution viscosity and affects the quality of electro nickel products. At the same time, unstable factors such as fine precipitates (colloids) and solid suspended solids or impurity ions are produced into the oil-in-water or water-in-oil emulsion. This is called the third phase. It was produced with the change of pH during extraction, which affects the quality of nickel. For example, Luo Jun [3] summed up that the third phase is formed in the presence of solid suspended matter in the solution, the following effect, like the formation of precipitation (or micelle) in the extraction, the degradation of extractant, and the presence of humic acid in the leaching solution (microorganisms or hydrolytic precipitation of metal ions such as Al, Fe) in the process of extraction and separation. The third phase increases the difficulty of the subsequent separation process. Moreover, the oil droplets dispersed in the nickel electrolyte are emulsified and dispersed in the whole solution or combined into flocs in the third phase with fine solids as stabilizers. The amount of the third phase and the elemental iron also increase, when the iron ion content in the nickel electrolyte increases [5]. Fine precipitates (colloids) are produced due to the change of pH or redox reaction in the process of extraction. The smaller particles are used as stabilizers to produce more stable emulsions [3, 6–10]. Liu Ying [11] et al. studied the extraction of Cerium by Bis(2-ethylhexyl) phosphate (DEHPA). It was found that DEHPA could reduce Ce^{4+} to form fine precipitates.

Therefore, it is necessary to analyze diagrams such as Fe-H_2O and Fe-S–Si-H_2O in sulfuric acid system in order to find out the cause of the formation of the third phase in the nickel electrolyte and improve the product quality. In this paper, the stable region of related iron compounds was obtained by thermodynamic analysis. The formation mechanism of the third phase was further proved by SEM–EDS, XRD, and infrared spectrum analysis.

Thermodynamic Analysis

The Eh–pH diagrams of Fe-H_2O, Fe-S-H_2O, and Fe-S–Si-H_2O are drawn by HSC Chemistry6.0.

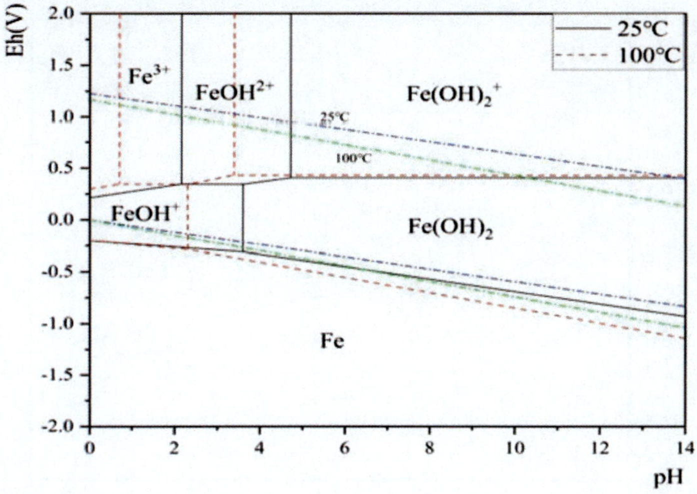

Fig. 1 Fe-H_2O system. (Color figure online)

Fe-H_2O System

Figure 1 shows iron exists in the oxidation state of III valence at high potential when the pH value is lower than 2.16 at 25 °C. The third valence iron begins to hydrolyze into $FeOH^{2+}$, when the pH value $\geq$ 2.16. The third valence iron continues to hydrolyze and synthesize $Fe(OH)_2^+$complex with the increase of pH value. The second valence iron is hydrolyzed to form $FeOH^+$ with the potential $\leq$ 0.21 V. It is the stable region of $FeOH^+$ when the potential value is between 0.20 and 0.21 and pH value is less than 3.6. The II valence molten iron is hydrolyzed into $Fe(OH)_2$ on the condition of pH value $\geq$ 3.6. Both II-valent iron complexes and iron compounds are reduced to elemental iron with the potential ≤ -0.20 V. The thermodynamic stability region of Fe^{3+} shrinks, the reaction equilibrium potential condition is basically unchanged, and the pH value of reaction equilibrium decreases at the temperature of the system rising to 100 °C. While the thermodynamic stability region of $FeOH^{2+}$ and $Fe(OH)_2^+$ moves to the left and the stable region of $Fe(OH)_2^+$ expands obviously. So the thermodynamic stability region of $Fe(OH)_2^+$ transition $Fe(OH)_2$ in the system is enlarged and the pH value of reaction equilibrium decreases.

Fe-S-H_2O System

Figure 2 shows iron exists in the form of Fe^{3+} under the condition of high potential value and strong acid at 25 °C. Fe^{3+} is hydrolyzed to $Fe(OH)_3$ at pH value $\geq$ 1.34. Fe^{3+} and $Fe(OH)_3$ are reduced to Fe^{2+} and $Fe(OH)_2$, respectively, with the decrease of potential. FeS is formed with potential value continues to decrease, because the

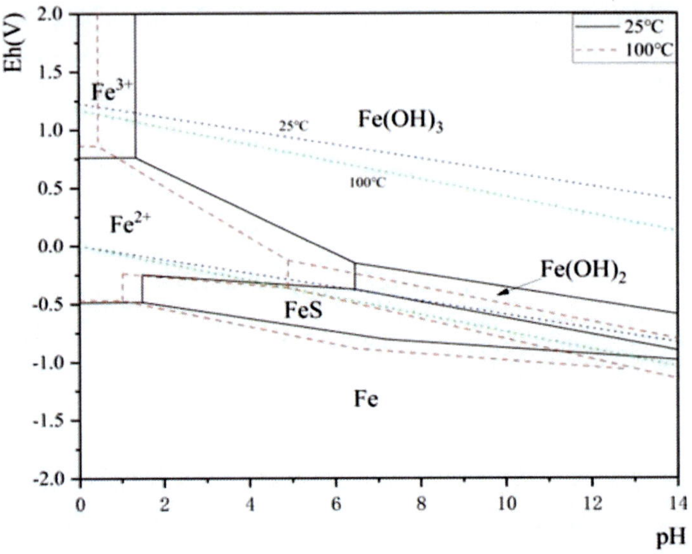

Fig. 2 Fe-S-H_2O system. (Color figure online)

affinity between iron and sulfur increases. The ions of II-valent iron ions and iron sulfides are reduced to elemental iron, when the potential ≤ -0.48 V. The thermodynamic stable region of Fe^{3+} and Fe^{2+} decreases at 100 °C. The reaction equilibrium potential changes little and the pH value of the reaction equilibrium decreases. The thermodynamic stable region of $Fe(OH)_3$ expands. So it is easier to generate $Fe(OH)_3$ in the system. The thermodynamic stability region of $Fe(OH)_2$ moves to the lower left and expands.

Fe-S-Si-H_2O System

Figure 3 shows the Eh–pH diagram of the Fe-S–Si-H_2O system. The thermodynamic stability region changes of Fe^{3+}, $FeOH^{2+}$, and $Fe(OH)_2{}^+$ are similar to those in this system in Fig. 1. While the potential ≤ -0.10 V and pH value is between 6.24 and 13.00 at 25 °C, the reduction is Fe_2SiO_4.$Fe(OH)_2$ was hydrolyzed and its stable region was small in the strong alkali environment with pH higher than 13.00. The thermodynamic stable region of Fe_2SiO_4 expands obviously and the stable region of $Fe(OH)_2$ increases slightly at 100 °C, while the stable region of $FeOH^+$ decreases obviously. $FeOH^+$, Fe_2SiO_4, and $Fe(OH)_2$ are reduced to elemental iron at the potential ≤ -0.19 V.

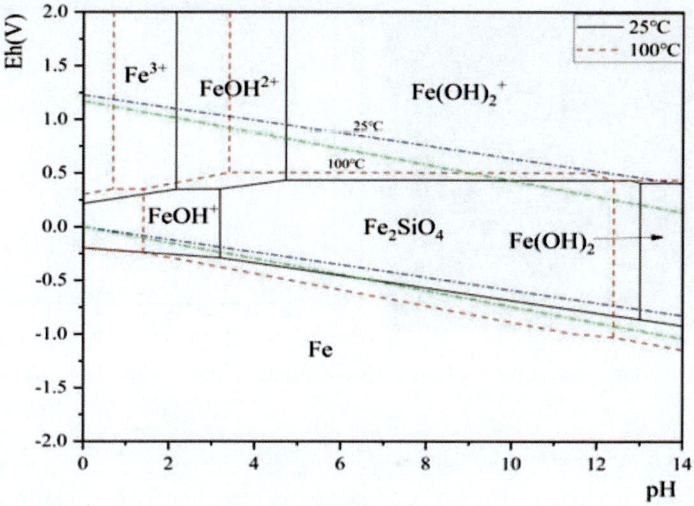

Fig. 3 Fe-S-Si-H_2O system. (Color figure online)

Results and Discussion

The third phase residue filtered from the extraction and the stripping is provided by a Chinese company.

The residue of the third phase filtered from extraction is tested by SEM–EDS. The results are shown in Table 1 and Fig. 4.

Table 1 Point A components of the third phase residue from extraction (Wt%)

Element	O	Na	Al	S	Fe	Ni
Content	5.69	1.47	1.46	23.06	62.97	5.35

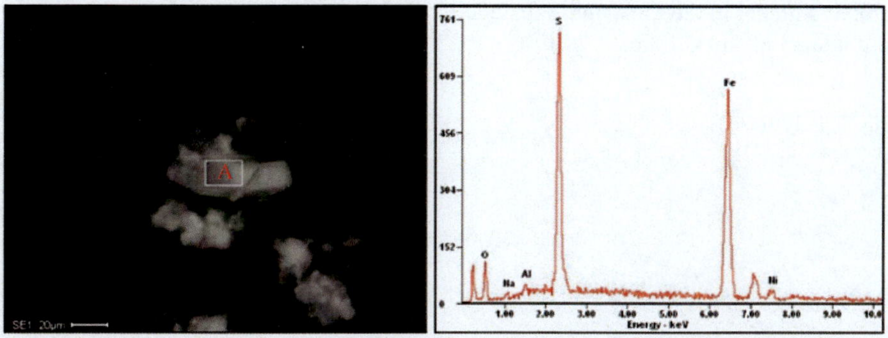

Fig. 4 SEM-EDS diagram of the third phase residue from extraction. (Color figure online)

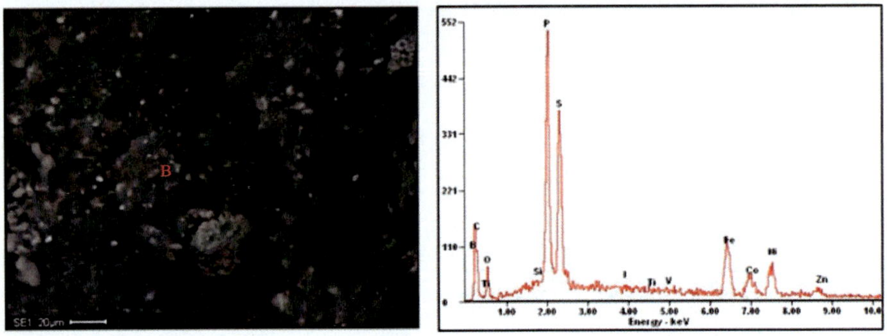

Fig. 5 SEM-EDS diagram of the third phase residue from stripping. (Color figure online)

Table 2 Point B components of the third phase residue from stripping (Wt%)

Element	B	C	O	Si	P	S	Co	Fe	Ni	Zn
Content	56.71	18.37	1.58	0.17	5.06	3.77	2.21	4.82	4.32	1.84

The residue of the third phase filtered from stripping was detected by SEM-EDS, and the results are shown in Fig. 5.

It can be seen from Table 2 that the main elements in the third phase residue from stripping are B, C, P, Fe, Ni, and so on. P from the extractant Cyanex272, and B from the boric acid added during extraction.

The flocs in the third phase are mainly suspended solids and charged colloids. Hydrated silica can be formed if a few solid particles of silica or silicate are dissolved in the nickel electrolyte. SiO_2 sol and gel can be synthesized by convergence after saturation due to the small solubility of hydrated silica. They are polycondensation and polymerization of silanol with silanol, silanol, and silyl ester to form a longer silicone chain in Fig. 6 [12, 13]. Silica gel particles will also increase the viscosity of nickel electrolyte, making the two phases more difficult to separate.

Combined with the results of XRD analysis in Fig. 7, it can be concluded that iron and nickel in the third phase residue from stripping are mainly formed by ferric sulfate and nickel sulfate.

Fig. 6 Polycondensation reaction [12, 13]

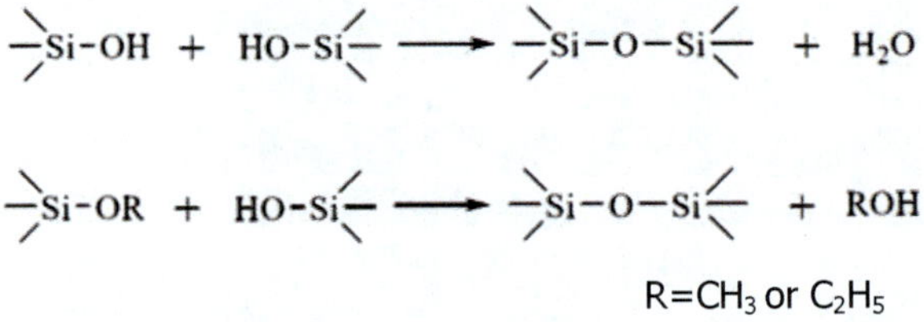

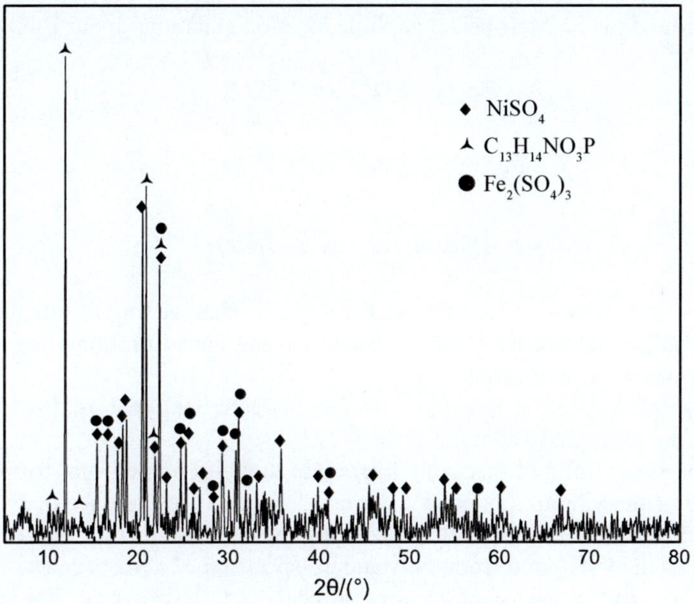

Fig. 7 XRD analysis of the third phase residue from stripping

The specific surface area of the third phase filter residue was measured in the extraction solution. The specific surface area obtained is 8.317 m^2/g. The third phase residue from extraction is detected by infrared spectrum. The result is shown in Fig. 8.

It can be seen from Fig. 8 that the four peaks at 1640, 1150, 1100, and 633 cm^{-1} are the main characteristic peaks of ferric sulfate hydrate. Therefore, the main substance in the third phase is ferric sulfate hydrate. Iron ions will be hydrolyzed to form Fe-SO_4

Fig. 8 Infrared spectrum diagram of the third phase residue from extraction. (Color figure online)

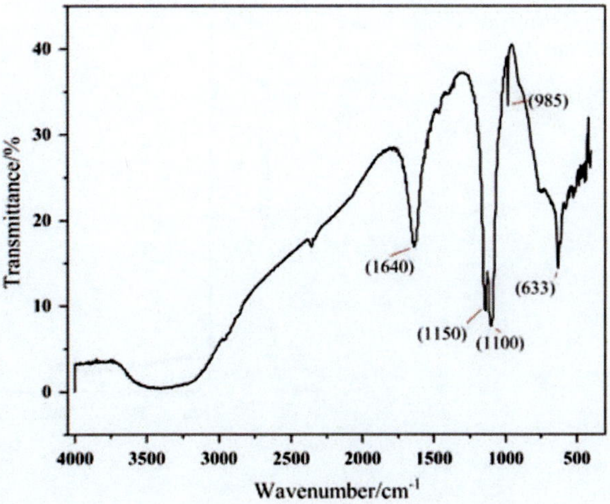

complex around pH = 2.5 [4]. The possible reaction equations are as follows:

$$Fe^{3+} + SO_4^{2-} \Leftrightarrow FeSO_4^+ \quad (1)$$

$$FeSO_4^+ + SO_4^{2-} \Leftrightarrow Fe(SO_4)_2^- \quad (2)$$

$$Fe^{3+} + HSO_4^- \Leftrightarrow FeHSO_4^{2+} \quad (3)$$

These Fe-SO_4 complexes can adsorb the opposite charged ions during the extraction process. The third phase can be formed by using the adsorption bridging effect of the polymer in the extractant.

The absorption peak at 985 cm^{-1} is the characteristic peak of P = O bond. Therefore, the third phase contains Cyanex272.

Figure 9 shows infrared spectrum diagram of the third phase from stripping. The occurrence of the vibration band at 1109 cm^{-1} implies the presence of Fe-OH-Fe. It can be seen that the two characteristic bands appear at 1470 cm^{-1} and 1377 cm^{-1} attributing to the C-H bond from the standard spectrum of kerosene. The vibration band at 1643 cm^{-1} appears the absorption peak of O-H in water. The stretching vibration in the 500–650 cm^{-1} region illustrates the presence of Fe-O bond. Iron ions will be hydrolyzed to form polynuclear hydroxyl complex ions around pH = 2.5 in nickel electrolyte [4]. Possible reaction equations are as follows:

$$Fe^{3+} + H_2O \Leftrightarrow FeOH^{2+} + H^+ \quad (4)$$

$$Fe^{3+} + 2H_2O \Leftrightarrow Fe(OH)_2^+ + 2H^+ \quad (5)$$

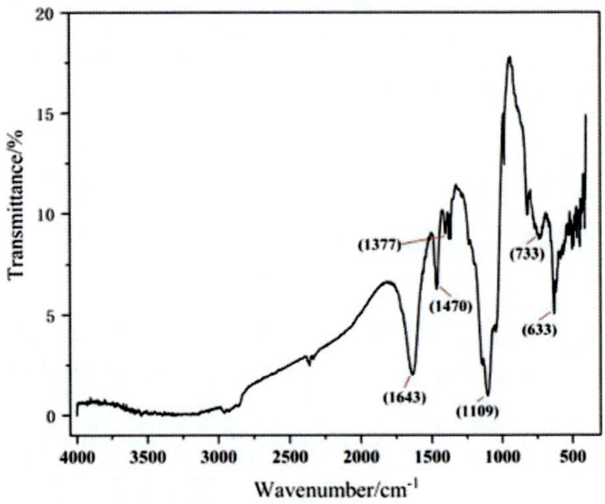

Fig. 9 Infrared spectrum diagram of the third phase from stripping. (Color figure online)

$$2FeOH^{2+} \Leftrightarrow Fe_2(OH)_2^{4+} \tag{6}$$

The products of Chinese formula (4) and formula (6) may also have a polymerization reaction:

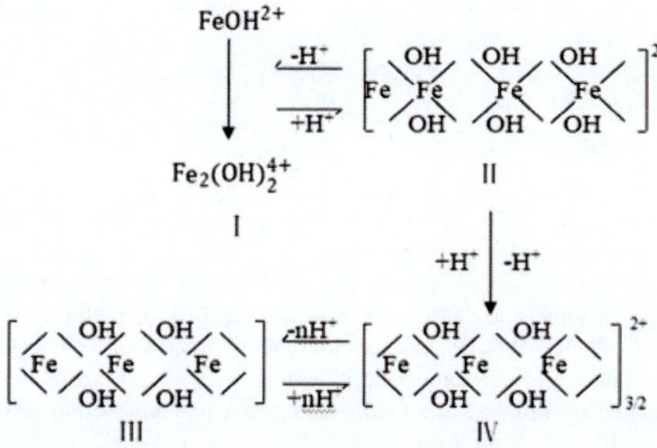

These multinuclear hydroxyl complex ions of iron can adsorb the opposite charged ions during the extraction process. They can make use of the adsorption bridging effect of the polymer in the extractant to form the third phase. Meanwhile, iron ions can be hydrolyzed to form ferric hydroxide colloid. The colloidal particles will aggregate into polymers, which increases the viscosity of the nickel electrolyte and promote the formation of the emulsion in the nickel electrolyte.

Conclusion

(1) The Eh-pH diagram of Fe-H_2O system shows that the thermodynamic stability region of $FeOH^{2+}$ and $Fe(OH)_2{}^+$ increases with the increase of temperature. This indicates that the increase of temperature is beneficial to the formation of iron polynuclear hydroxyl complex ions. The Eh-pH diagram of Fe-S-H_2O system shows that III valent iron begins to hydrolyze to form $Fe(OH)_3$ colloid at pH = 1.34. Furthermore, the colloidal particles will aggregate into polymers. The thermodynamic stability region expands with the temperature increases to 100 °C, but the pH value decreases to 0.45. So the hydrolysis of iron to polynuclear hydroxyl complex ions or Fe $(OH)_3$ colloid will affect the formation of the third phase.

(2) The thermodynamic stability regions of Fe^{3+}, $FeOH^{2+}$, and $Fe(OH)_2{}^+$ are basically the same when S and Si are introduced into Fe-H_2O system. Therefore, the introduction of S and Si will hardly affect the complexes of III valence iron series. The thermodynamic stability region of the complex $FeOH^+$ of II valent iron is slightly reduced, and the equilibrium pH moves to the left.

(3) The iron ion will form multinuclear hydroxyl complex ion or Fe-SO_4 complex after hydrolysis and complexation in nickel electrolyte, according to the analysis of SEM-EDS and infrared spectrum. The complex can form the third phase, because the oppositely charged ions were adsorbed and it makes use of the adsorption bridging effect of the polymer in the extractant.

References

1. Zhang QX, Zhang GQ, Tang RR, et al (2014) Principles and practice of extractive metallurgy M. Central South University Press, Changsha
2. Lu JB, Feng JH, Zheng JF, Liu GL, Guo Y, Li YJ (2016) Application of modified carbon materials in the purification and degreasing of nickel electrolyte. Chem Eng 44(10):9–13
3. Luo J, Xie T, Yao G, et al (2014) The causes and treatment methods of three-phase extraction in metallurgy: Chinese academy of engineering% Chinese society of metals. In: The 8th national academic conference on mineral processing and green and high-efficiency development of mineral resources utilizing summit forum proceedings, pp 433–435
4. Liu J, Fang JW, Qiu GZ, et al (2000) Study on the formation mechanism and control methods of the third phase in the process of copper extraction. Copp Eng 2:29–33
5. Wu QL, Liu SP (2017) Study on the formation and prevention of the third phase in the process of extraction. World Nonferrous Metals 21:10–10
6. Liang EL, Sangeetaprivya PS, Yong KH, Eng SC, Beng TT (2020) Recent advances of characterization techniques for the formation, physical properties and stability of pickering emulsion. Adv Colloid Interface Sci 277, 102117
7. Wu J, Ma GH (2016) Recent studies of Pickering emulsions: particles make the difference. Small 12:4633–4648
8. Li C, Li Y, Sun P, Yang C (2013) Pickering emulsions stabilized by native starch granules. Colloids Surf A Physicochem Eng Asp 431:142–149
9. Tambe DE, Sharma MM (1994) The effect of colloidal particles on fluid-fluid interfacial properties and emulsion stability. Adv Colloid Interface Sci 52:1–63
10. Varanasi S, Henzel L, Mendoza L, Prathapan R, Batchelor W, Tabor R, et al (2018) Pickering emulsions electrostatically stabilized by cellulose nanocrystals. Front Chem 6:1–9
11. Liu Y, Long ZQ, Huang, WM, et al (2001) The reason for the formation of the third phase in the process of extracting cerium from fluorine-containing rare earth sulfate solution. Chin J Rare Earths, Jing 19(4):320–323. https://doi.org/10.3321/j.issn:1000-4343.2001.04.007
12. Liu YH (2019) Regulation of structure and surface properties of silica gel and its application in typical VOC purification. Hebei Normal University, Hebei
13. Li L, Ruan RM, Wen JK, et al (2008) Analysis of the effect of solid particles on the formation of the third phase in the process of copper extraction. Rare Met 32(6):748–753

Phase Diagram and Thermodynamic Properties of Cu–O Binary System

Shadia J. Ikhmayies

Abstract Thermo-Calc software is used to reassess the equilibrium temperature-composition phase diagram at 1 atm and to deduce the thermodynamic properties of Cu–O binary system at T = 1500 K. The phase diagram showed one single phase which is the non-stoichiometric Cu ionic liquid other than elemental Cu and O at the terminals, in addition to Cu_2O and CuO which can be considered as stoichiometric line compounds. There are nine fields of two mixed phases, two eutectic reactions, one monotectic reaction, and a miscibility gap. The solubility of oxygen in copper and that of copper in oxygen are determined. The molar Gibbs energy and molar enthalpy curves of the Cu_2O–CuO ionic solution are plotted against mole fraction oxygen at 1500 K. The natural logarithm of the activity is plotted against oxygen mole fraction for oxygen and copper in the melt, and the activity coefficients are determined.

Keywords Phase diagram · Thermodynamic properties · Calphad · Copper oxides

Introduction

The Cu–O system contains two stable oxides, cupric oxide (CuO) and cuprous oxide (Cu_2O), both with direct band gaps. Copper oxides are among the most important transition-metal oxides, where both CuO and Cu_2O are used as p-type semiconductors, high critical temperature (T_C) superconductors, and magnetic storage media. In addition, they are used for gas sensing, in catalysis, and battery applications. This is due to their low bandgap energy, which is 1.2 eV for CuO and 2.2 eV for Cu_2O, high optical absorption, nontoxic nature, and the economics of these materials [1–3]. The band gap of these copper oxides can be tuned between 1.6 eV for CuO to 2.3 eV for Cu_2O, respectively. Copper oxide CuO has attracted attention because it is the simplest member of a family of compounds, which exhibit certain physical

S. J. Ikhmayies (✉)
Jabal El-Hussain, Amman 11121, Jordan
e-mail: shadia_ikhmayies@yahoo.com

B. Li et al. (eds.), *Materials Engineering—From Ideas to Practice: An EPD Symposium in Honor of Jiann-Yang Hwang*, The Minerals, Metals & Materials Series,
https://doi.org/10.1007/978-3-030-65241-8_13

properties such as high-temperature superconductivity, electron correlation effects, and spin dynamics [4, 5].

In this work a brief reassessment of the phase diagram and thermodynamic properties of the Cu–O system at atmospheric pressure was performed using Thermo-Calc 2020b software. The phase diagram was discussed, and the invariant reactions, melting temperatures, and solubility of Cu in O and of O in Cu were determined. The molar Gibbs energy, molar enthalpy, and activities of the components were all determined at T = 1500 K.

Methodology

Thermo-Calc software-2020b is used to perform different kinds of thermodynamic and phase diagram calculations in equilibrium problems by the Calphad method [6], where this word was derived from the phrase (**CAL**culation of **PHA**se **D**iagrams). The software is based upon a powerful Gibbs Energy Minimizer that utilizes Gibbs free energy minimization procedure to calculate phase equilibria and thermodynamic properties of a chosen system. The Calphad method is based on deriving the thermodynamic functions of a system from all available experimental data, where the thermodynamic functions are represented as polynomials of chemical composition and temperature. Then, the values of the polynomial coefficients are obtained using numerical optimization techniques [7, 8]. The Thermo-Calc software has two main components: the application itself, and the internally consistent thermodynamic dataset it uses. For a given set of conditions, the computer determines the change in free energy for each possible combination of phases and phase compositions. Then, it selects the state that minimizes the total Gibbs free energy [9].

In this work as in previous works [10–16], the template "Binary Calculation" with the type of calculation "phase diagram " was chosen to calculate the equilibrium phase diagram of the Cu–O binary system, and the type "Gibbs energy curves" to determine the thermodynamic properties such as Gibbs energy, enthalpy, and activity curves. The total pressure is 1 atm, and for the phase diagram , the temperature range was automatically selected, and just stable phases were credited. The used database for all plots is the TCBIN: TC Binary Solutions v1.1 database.

Results and Discussion

Phase Diagram

Figure 1 displays the temperature-composition phase diagram of the Cu–O binary system calculated at 1 atm, where the composition is represented by mole fraction O (X_O). From the figure, there is one single phase other than elemental

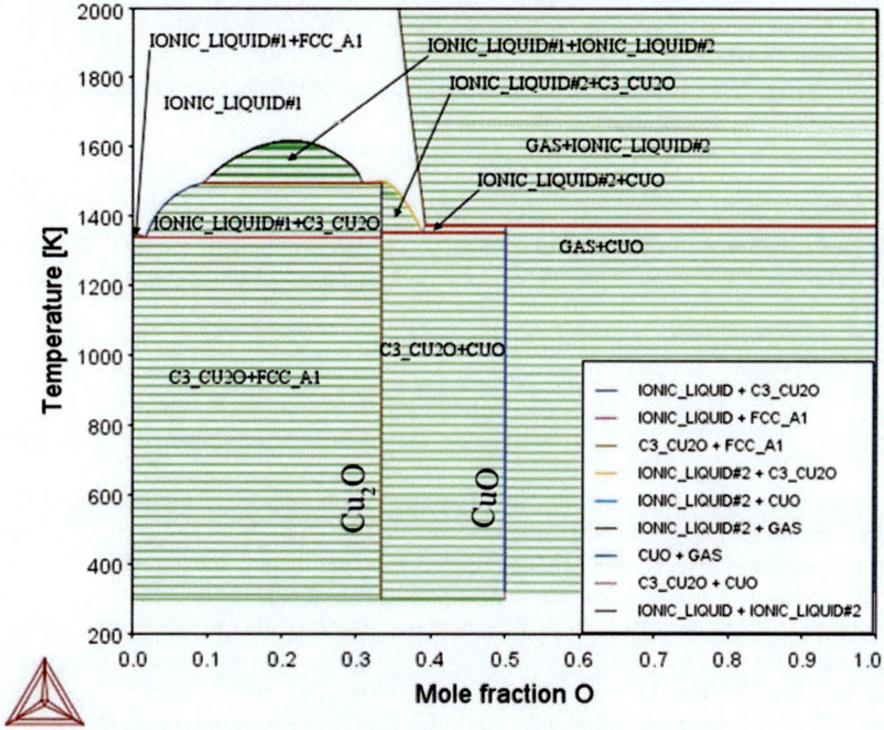

Fig. 1 Phase diagram of the Cu–O system obtained in this work. Note: The labels CuO and Cu_2O are not produced by Thermo-Calc 2020b software, but inserted by the author. (Color figure online)

Cu and O at the terminals, which is the Cu ionic liquid 1 (IONIC_LIQUID#1). Besides, the phase diagram in Fig. 1 shows that CuO and Cu_2O can be considered as line compounds, but in Ref. [17] they are indicated as phases of fixed composition with negligible departures from stoichiometry. In addition, the figure shows nine fields of two mixed phases which are Cu–Cu_2O (C3_CU2O + FCC_A1), Cu ionic liquid-Cu (IONIC_LIQUID#1 + FCC_A1), Cu ionic liquid-Cu_2O (IONIC_LIQUID#1 + C3_CU2O), Cu ionic liquid- Cu_2O ionic liquid (IONIC_LIQUID#1 + IONIC_LIQUID#2), Cu_2O–CuO (C3_CU2O + CUO), Cu_2O ionic liquid-Cu_2O (IONIC_LIQUID#2 + C3_CU2O), Cu_2O ionic liquid- CuO (IONIC_LIQUID#2 + CUO), gas- CuO (GAS + CUO), and CuO ionic liquid-gas (GAS + IONIC_LIQUID#2). Using the two sublattice model for ionic liquids developed by Hillert et al. [18, 19], the Cu melt (Ionic Liquid 1) consists of: the two sublattices (Cu^{+1}, Cu^{+2}), (O^{-2}, VA, O), and Cu_2O melt (IONIC_LIQUID#2) consists of the two sublattices (Cu^{+1}, Cu^{+2}), (VA, O^{-2}, O). Liquid 1 is Cu rich while liquid 2 is O rich. The gas consists of Cu, CuO, O, O_2, and O_3.

The melting temperature of FCC–Cu is 1357.23 K, while the accepted value is 1357.6 K [20]. The congruent melting temperature of Cu_2O is 1497.26 K, while Slade and Farrow [21] gave a value of 1483 K, and Taskinen [22] gave a value of

1500 K. The congruent melting point of CuO is 1373.88 K from Fig. 1, but the experimental value given by Šesták [17] is about 1110 °C (1383 K).

From Fig. 1, the Cu–Cu_2O eutectic point is found at 0.01725 mol fraction O and T = 1339.29 K, where the reaction is

$$\text{Liqud 1} \rightarrow \text{Cu} + \text{Cu}_2\text{O} \quad (1)$$

This is larger than the value 1338 K at Xo = 0.0154 obtained by Heyn [23], and approximately the same as the value 1339.6 K at Xo = 0.0174 found in Taskinen [24].

The Cu_2O-CuO eutectic from Fig. 1 is found at 0.388 mol fraction O and T = 1353.24 K, and it is given by the reaction

$$\text{Liqud 2} \rightarrow \text{Cu}_2\text{O} + \text{CuO} \quad (2)$$

This position is consistent with the experimental one given by Smyth and Roberts [25, 26] which is 1353 K at Xo = 0.389, and that given by Boudène et al. [27, 28], which is 1353 K and Xo = 0.388, and finally the experimental and assessed value of T = 1354 K at 0.388 mol fraction O given by Hallstedt et al. [29].

There is a miscibility gap located in the range 0.094–0.308 mol fraction O, which starts at T = 1497.26 K and disappears at 1617.47 K. Between these two temperatures the two liquids (Liquid 1 and liquid 2) are immiscible or they don't mix. After 1617.47 K they become completely miscible. This position of the miscibility gap is consistent with that shown in Ref. [29] which is T = 1498 K and 0.092–0.311 mol fraction. Taskinen [22] got it at 1498 K and Xo = 0.096–0.309, but Slade and Farrow [21] got it at 1468 K and Xo = 0.084–0.319.

There is a monotectic reaction at 1497.26 K in the miscibility region, where

$$\text{Liqud 2} \rightarrow \text{Liquid 1} + \text{Cu}_2\text{O} \quad (3)$$

Figure 2 shows enlarged parts of Fig. 1 to estimate the solubility of O in solid Cu and of Cu in O, where Fig. 2a shows the O-rich side, and Fig. 2b shows the Cu-rich side. From Fig. 2a the solubility of solid Cu in oxygen is found to be 0.0008 mol fraction Cu at 2000 K or 0.08 at.% Cu. While the solubility of oxygen in solid Cu from Fig. 2b is 0.00022 mol fraction O or 0.022 at.% O at 1339.4 K.

Thermodynamic Properties

The molar Gibbs energy, molar enthalpy, and activities of components are all calculated relative to stable element reference (SER). Figure 3 shows the plot of the molar Gibbs energy against O mole fraction for the phase Cu_2O–CuO ionic solution (liquid 2) at 1500 K after the melting of all solids and the appearance of the miscibility gap.

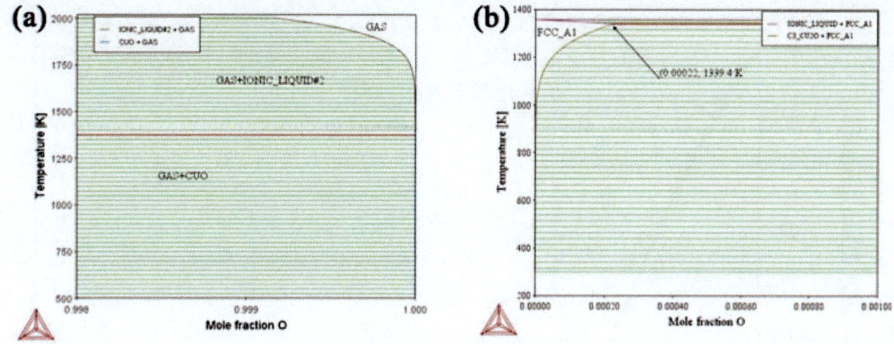

Fig. 2 Enlarged parts of Fig. 1 to find the solubilities. **a** O-rich side. **b** Cu-rich side. (Color figure online)

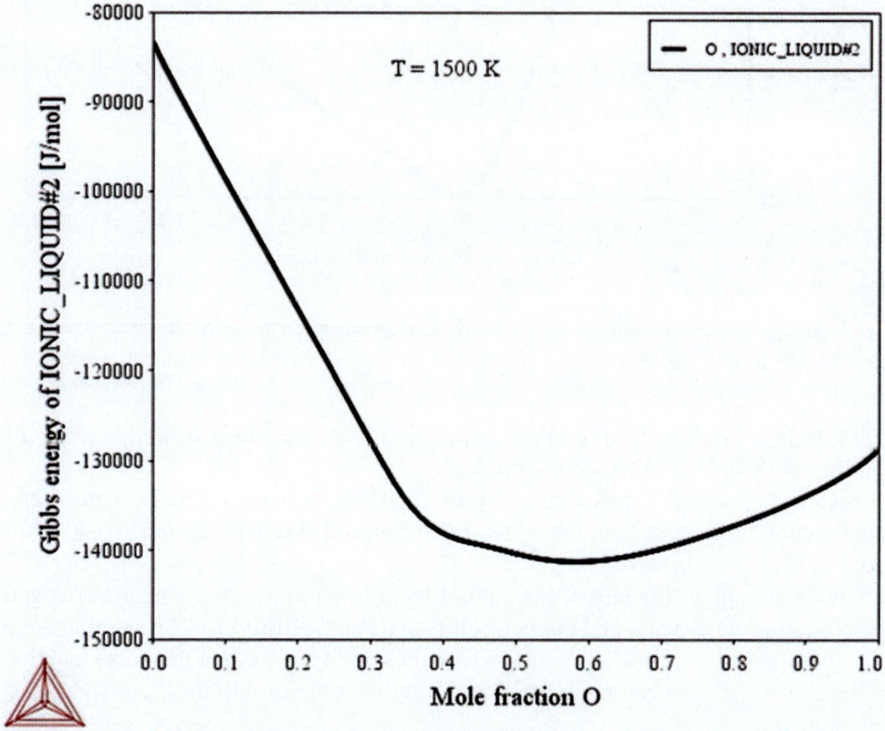

Fig. 3 Gibbs energy against O mole fraction for the Cu_2O–CuO ionic solution at 1500 K. (Color figure online)

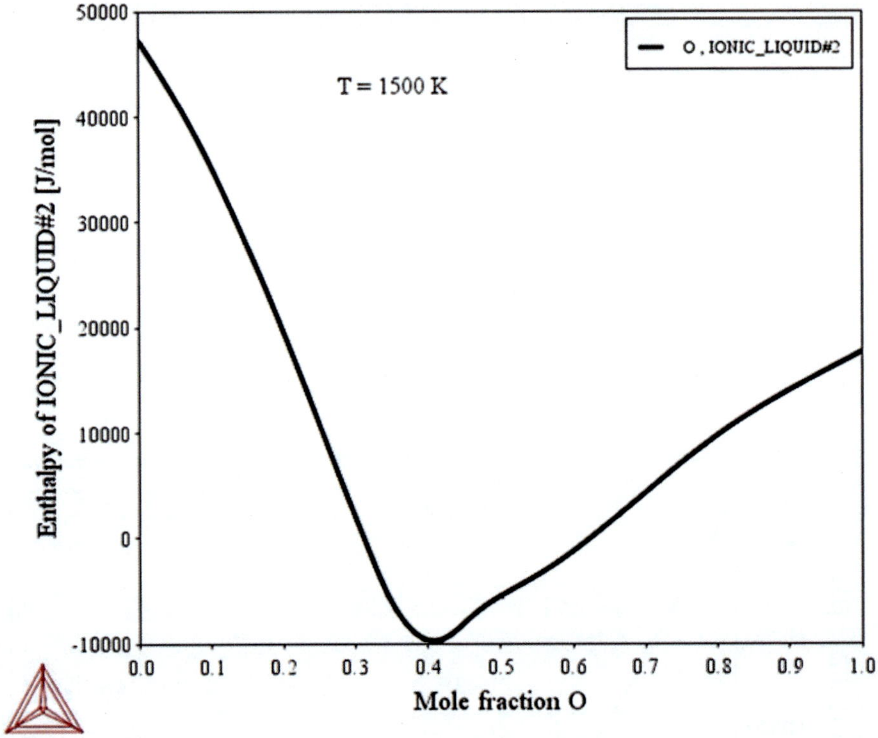

Fig. 4 The plot of molar enthalpy of the Cu_2O–CuO ionic solution against mole fraction O at 1500 K. (Color figure online)

The breadth of the curve around the minimum means that there is a large range of stability of this phase at this temperature.

Figure 4 shows the molar enthalpy of Cu_2O–CuO ionic solution at the same temperature. From the figure, the minimum enthalpy at the eutectic point at 0.407 mol fraction O is –10 kJ/mole.

Figure 5 displays the plot of the natural logarithm of the activity against oxygen mole fraction for oxygen and copper in the melts at T = 1500 K. The sharp change of activities occurs around $X_O = 0.5$, where the activity of oxygen increases and that of decreases with oxygen mole fraction. Using the relationship for the activity of a component i;

$$a_i = \gamma_i X_i \tag{4}$$

where γ_i is the activity coefficients.Taking the natural logarithm of both sides to get

$$\mathrm{Ln}(a_i) = \mathrm{Ln}(\gamma_i) + \mathrm{Ln}(X_i) \tag{5}$$

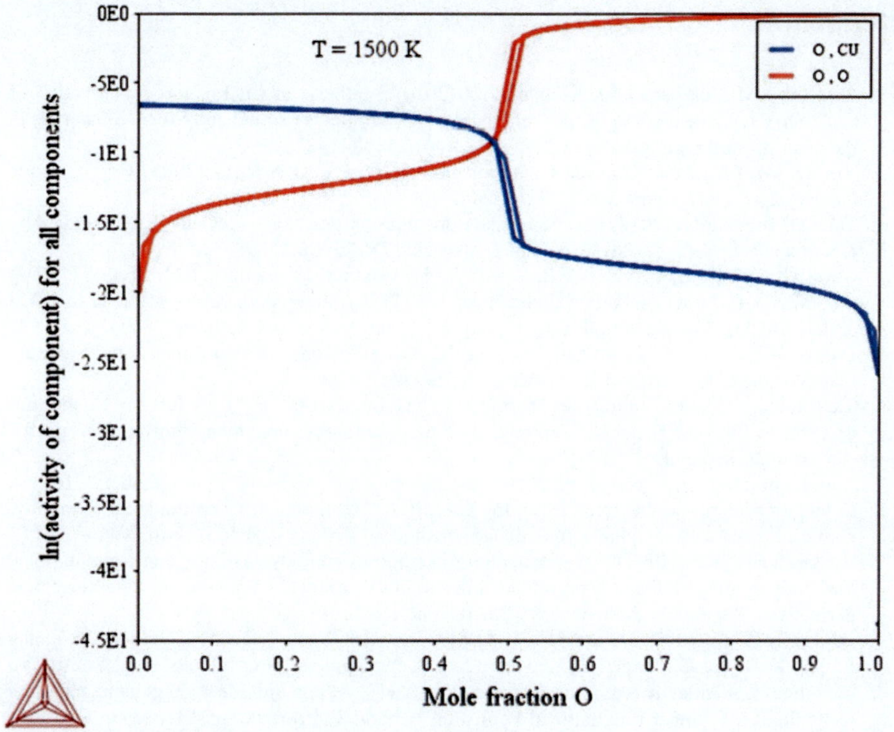

Fig. 5 The natural logarithm of the activity against oxygen mole fraction for oxygen and copper in the melts. (Color figure online)

So from Fig. 5 for $X_O = 1$, $Ln(a_O) = 0$, or $a_O = 1$, but for $X_O = 0$, $X_{Cu} = 1$, and $Ln(a_{Cu}) = Ln(\gamma_{Cu}) = -6.57$, then $\gamma_{Cu} = 0.0014$.

Conclusions

The phase diagram of Cu–O system was reassessed at 1 atm using Thermo-Calc 2020b and the single and mixed phases are identified. The melting points, invariant reactions, and solubilities are determined. The molar Gibbs energy and molar enthalpy curves were plotted. A semi-logarithmic plot of the activity against mole fraction O was performed and activity coefficients were determined. These results are important for the different applications of CuO and Cu_2O.

References

1. Shrestha KM, Sorensen CM, Klabunde KJ (2010) Synthesis of CuO nanorods, reduction of CuO into Cu nanorods, and diffuse reflectance measurements of CuO and Cu nanomaterials in the near infrared region. J Phys Chem C 114(34):14368–14376
2. Yin M, Wu CK, Lou Y, Burda C, Koberstein JT, Zhu Y, O'Brien S (2005) Copper oxide nanocrystals. J Am Chem Soc 127(26):9506–9511
3. Tahir D, Tougaard S (2012) Electronic and optical properties of Cu, CuO and Cu_2O studied by electron spectroscopy. J Phys Condens Matter 24(17):175002
4. Cava RJ, Batlogg B, Vandover RB, Krajewski JJ, Waszcak JV, Fleming RM, Peck WF, Rupp LW, Marsh P, James ACWP, Schneemeyer LF (1990) Superconductivity at 60-K in LA2-XSRXCACU2O6—the simplest double-layer cuprate. Nature 345(6276):602–604
5. Dagotto E, Rice TM (1996) Surprises on the way from one- to two-dimensional quantum magnets: the ladder materials. Science 271(5249):618–623
6. Thermo-calc documentation set thermo-calc version 2017a. In: Introduction to thermo-calc, https://www.thermocalc.com/media/40962/thermo-calc-documentation-set.pdf, p. 97. Accessed 26 September 2017
7. Introduction to thermo-calc. https://www.thermocalc.com. Accessed 18 September 2016
8. Calculation of phase diagrams using the CALPHAD method. In: Computational thermodynamics. https://www.calphad.com/calphad_method.html. Accessed 16 August 2019
9. Boutwell BA et al. (1997)Phase formation modeling of an alloy casting using computational thermodynamics. In: Paper is presented at the 4th international symposium on superalloys and derivatives, Pittsburgh, Pennsylvania, 15–18 June 1997
10. Ikhmayies SJ (2019) Phase diagram of Al-Si system. In: Wang T, Chen X, Guillen DP, Zhang L, Sun Z, Wang C, Haque N, Howarter JA, Neelameggham NR, Ikhmayies S, Smith YR, Tafaghodi L, Pandey A (eds) Energy technology 2019: carbon dioxide management and other technologies. Springer International Publishing, Cham, Switzerland, pp 231–237
11. Ikhmayies SJ (2018) Thermo-calc of the phase diagram of the Fe–Si system. In: Sun Z, Wang C, Guillen DP, Neelameggham NR, Zhang L, HowarterJA, Wang T, Olivetti E, Zhang M, Verhulst D, Guan X, Anderson A, Ikhmayies S, Smith YR, Pandey A, Pisupati SV, Lu H (eds) Energy technology 2018: carbon dioxide management and other technologies. Springer International Publishing, Cham, Switzerland, pp 471–477
12. Ikhmayies SJ (2018) Thermo-calc of the phase diagram of calcium silicon (Ca–Si) System. In: Sun Z, Wang C, Guillen DP, Neelameggham NR, Zhang L, HowarterJA, Wang T, Olivetti E, Zhang M, Verhulst D, Guan X, Anderson A, Ikhmayies S, Smith YR, Pandey A, Pisupati SV, Lu H (eds.) Energy technology 2018: carbon dioxide management and other technologies. Springer International Publishing, Cham, Switzerland, pp 489–494
13. Ikhmayies SJ (2017) Phase analysis of the Si–O_2 system. In: Zhang L, DrelichJW, Neelameggham NR, Guillen DP, Haque N, Zhu J, Sun Z, Wang T, HowarterJA, Tesfaye F, Ikhmayies S, Olivetti E, Kennedy MW (eds) Energy technology 2017: carbon dioxide management and other technologies. Springer International Publishing, Cham, Switzerland, pp 333–342
14. Ikhmayies SJ (2020) Phase diagram of In-P binary system. In: Li J, Zhang M, Li B, Monteiro SN, Ikhmayies SJ, Kalay YE, Hwang J, Escobedo-Diaz JP, Carpenter JS, Brown AD (eds) Characterization of minerals, metals and materials 2020. Springer International Publishing Cham, Switzerland, pp 283–288
15. Ikhmayies SJ (2018) Thermo-calc of the phase diagrams of the Nb-N system. In: TMS 2018 147th annual meeting & exhibition supplemental proceedings. Springer International Publishing. Switzerland, Cham, pp 755–760
16. Ikhmayies SJ (2020) Using thermo-calc software to produce the phase diagram of Zn–Te system. J Energy Syst 4(3):88–95
17. Šesták J (1992) Phase diagrams in CuO_x, based superconductors. Pure & Appl Chem 64(1):125–136
18. Hillert M, Jansson B, Sundman B, Ågren J (1985) A two-sublattice model for molten solutions with different tendency for ionization. Metall Trans A 16(2):261–266

19. Sundman B (1991) Modification of the two-sublattice model for liquids. Calphad 15(2):109–119
20. Melting points of elements reference. Angstromsciences. https://www.angstromsciences.com/melting-points-of-elements-reference. Accessed 12 October 2020
21. Slade RE, Farrow ED (1912) An investigation of the dissociation pressures and melting points of the system copper-cuprous oxide. Proc R Soc (London) A 87:524–534
22. Taskinen P (1984) Thermodynamics of liquid copper-oxygen alloys at 1065–1450 °C. Scand J Metall 13:75–82
23. Heyn E (1904) Kupfer und Sauerstoff. Z Anorg Chem 39:1–23
24. Taskinen P (1981) Liquidus equilibria and solution thermodynamics in copper-rich copper-nickel-oxygen alloys. Acta Polytechnica Scand. Chem Incl Metall Ser. Ch 145:1–45
25. Smyth EH, Roberts HS (1920) The system cupric oxide, cuprous oxide, oxygen. J Am Chem Soc 42:2582–2607
26. Roberts HS, Smyth EH (1921) The system copper: cupric oxide: oxygen. J Am Chem Soc 43:1061–1079
27. Boudène, A (1991) Thermochemical investigations in the system La-Sr-Cu-O. PhD Thesis, T.H. Aachen (in German)
28. Boudène A, Hack K, Mohammad A, Neuschiitz D, Zimmermann E (1992) Experimental investigation and thermochemical assessment of the system Cu–O. Z Metallkd 83(9):663–668
29. Hallstedt B, Risold D, Gauckler LJ (1994) Thermodynamic assessment of the copper-oxygen system. J Phase Equilib 15(5):483–499

A Case Study of Sintering with Low Silica Iron Ore

Liangping Xu, Huibo Liu, Xiduan Yang, Hao Yin, Mingjun Rao, Qiang Zhong, Guanghui Li, and Tao Jiang

Abstract The low-silica and low basicity sintering is an effective measure for improving iron grade of sinter and reducing the slag amount of ironmaking and steelmaking. Sintering of the mixture at basicity (1.9, 1.8, and 1.7) and SiO_2 content (5.0, 4.9, 4.8, 4.7, and 4.6) were conducted in a sintering pot, and the microstructures of typical sinters were taken into comparison in this work. Experimental results show that with the basicity decreases from 1.90 to 1.70, there is an obvious change in the tumbler strength index (TI) of sinter; SiO_2 content affects mainly on the yield, but to a lesser extent on the TI of sinter. At the basicity of 1.70 and SiO_2 content of 4.60%, the TI of sinter is 58.53% which is 5% lower than the sinter at basicity of 1.90 and SiO_2 content of 4.98%. At basicity of 1.80 and SiO_2 content of 4.60%, the sinter obtains a good performance (yield of 76.07% and TI of 60.09%), which is close to the indexes of sinter at relatively high-silica and high basicity (basicity of 1.90 and SiO_2 content of 4.98%), indicating that low SiO_2 content and low basicity sintering can be implemented.

Keywords Iron ore sintering · Low SiO_2 content · Low basicity · Tumbler strength index

L. Xu · H. Liu · M. Rao (✉) · Q. Zhong (✉) · G. Li · T. Jiang
School of Minerals Processing and Bioengineering, Central South University, Changsha 410083, China
e-mail: mj.rao@csu.edu.cn

Q. Zhong
e-mail: zhongqiang2008csu@163.com

X. Yang · H. Yin
Hunan Valin Lianyuan Iron and Steel Co., Ltd., Loudi 417009, China

B. Li et al. (eds.), *Materials Engineering—From Ideas to Practice: An EPD Symposium in Honor of Jiann-Yang Hwang*, The Minerals, Metals & Materials Series,
https://doi.org/10.1007/978-3-030-65241-8_14

Introduction

Iron and steel are important materials that can be recycled and never lose their properties in construction, shipbuilding, military, and other industries [1]. According to the report from World Steel Association, about 1.9 billion tons of crude steel were produced in the world in 2019 [2]. Such a large output makes a difference in all aspects of human beings, such as the consumption of fuels and the emission of pollutants [3].

Improving the productivity of ironmaking is of great significant research direction for reducing pollution and fuel consumption [4], by means of improving the iron grade of raw materials, the productivity of agglomeration, and the working efficiency of blast furnace and converter [5]. In the process of iron and steel smelting, the productivity of converter depends on the quality of hot metal produced by the blast furnace . Furthermore, the quality of hot metal in blast furnace generally depends on the process of agglomeration, and the productivity of agglomeration is affected by the grade of iron ore, granulation, and sintering velocity [6]. However, improvement of iron grades in raw materials commonly depends on the concentration of iron ore, during which the gangue is separated from the iron-bearing minerals. But some impurities exist in the form of isomorphism or inclusion, which makes it difficult to separate. Such a tricky problem is left to the agglomeration process, in which different materials are proportioned to keep a certain chemical composition. Naturally, improving the iron grade of sinter becomes an important topic in ironmaking and steelmaking . Low SiO_2 content and low basicity sintering are thought as an effective measure to improve it. In the past, most of the researches focus on the impact of a single factor [7]. Wu et.al investigated the metallurgical performance of low silicon sinter and found that SiO_2 content decrease from 5.0 to 4.8%, the metallurgies performance had a little change [8]. Zhou et al. indicated that the SiO_2 content was not the lower the better, and an appropriate SiO_2 content depends on the raw material used in the plant [9].

In order to find the suitable SiO_2 content and basicity of Valin LY Steel and research the characteristic of low SiO_2 content and low basicity sintering, a series of tests including effect of SiO_2 content, effect of basicity , and comprehensive influence of SiO_2 content and basicity were carried out. The microstructure of sinter obtained from different SiO_2 content and basicity is researched.

Materials and Methods

Materials

Raw materials for sintering include iron ores, return fine (−5 mm sinter), coke breeze, and fluxes (quicklime, dolomite). The main chemical composition of raw materials shown in Table 1. A, B, C, D, E, F, and G are iron ores. It shows that iron content of B,

Table 1 Main chemical composition of raw materials (wt.%)

Materials		TFe	FeO	SiO_2	CaO	MgO	Al_2O_3	K_2O	Na_2O	P	S	LOI
Iron ore	A	59.91	2.51	7.39	2.23	0.49	0.55	0.09	0.09	0.50	0.05	1.37
	B	63.46	0.07	4.95	0.08	0.03	1.95	0.27	0.03	0.06	0.02	0.90
	C	65.94	1.22	1.31	0.05	0.05	1.38	0.01	0.01	0.05	0.02	2.76
	D	57.55	0.03	5.22	0.08	0.07	2.29	0.05	0.03	0.06	0.03	7.70
	E	62.37	25.09	10.54	0.16	0.30	0.93	0.04	0.03	0.04	0.06	0.97
	F	59.81	0.32	3.90	0.07	0.06	2.16	0.03	0.05	0.07	0.03	6.58
	G	68.24	19.12	3.37	0.31	0.35	0.37	0.02	0.04	0.01	0.04	0.41
Return fine		56.84	6.81	4.92	9.36	1.95	1.58	0.08	0.04	0.07	0.02	0.00
Coke breeze		0.81	–	6.04	0.49	0.12	2.21	0.09	0.07	0.12	0.03	88.08
Dolomite		–	–	2.13	0.24	31.90	19.41	0.06	0.03	0.01	0.00	44.64
Quicklime		–	–	1.66	0.10	79.95	3.08	0.09	0.02	0.01	0.05	14.70

LOI Loss on ignition

C, and G ore is higher than 63%; SiO_2 content of A, D, and E ore is higher than 5%, especially the SiO_2 content of E ore is 10.54%, which provides a good measure to realize low silicon and low basicity sintering. So E ore is used to reduce the content of SiO_2, while fluxes are used to adjusted basicity (CaO/SiO_2) and MgO content.

Methods

Experimental Procedure

The reference ratio of blended ore is 8% A, 13% B, 15% C, 6% D, 12% E, 37% F, and 9% G. In order to obtain a suitable SiO_2 content, the proportion of E ore in blended ore was reduced and the proportion of other iron ores in blended ore was fixed. In other words, the proportion of other ores increases proportionally.

The experimental procedure is a traditional iron ore sintering process, which is shown with a schematic diagram in Fig. 1. In this study, all of the experiments were carried out in a pilot-scale test with a 170 mm diameter sintering pot. To prepare sinter mixture, all of the iron ore, fluxes, return fines, and coke breeze are granulated in a drum mixer which is 1000 mm in length and 600 mm in diameter for 5 min. The sinter mixture was loaded into a sintering pot with 900 mm in height, and ignited at 1150 °C ± 50 °C with suction negative pressure of 5 kPa for 2 min. To ensure the surface mineralization effect, heat preservation is implemented above 900 °C with suction negative pressure of 5 kPa for 1 min. Afterward, suction negative pressure was promptly increased to 10 kPa. At the end of sintering, it needs 3 min to cool sinter at suction negative pressure 5 kPa. Total sintering time is measured from ignition to the point at which the flue gas reached its maximum temperature.

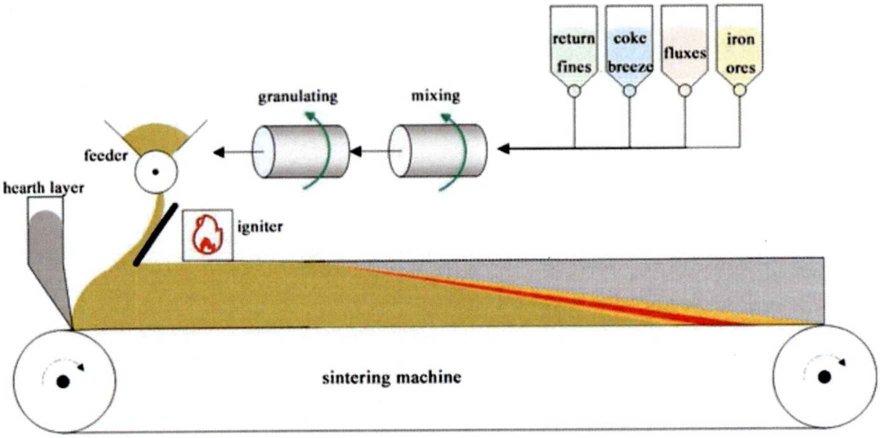

Fig. 1 Schematic of the sintering process. (Color figure online)

Sintering Indexes

Indexes involving sinter yield, tumbler index (TI), productivity (P), solid fuel consumption (SFC), Vertical sintering velocity (VSV), were set as important evaluation indexes to the sintering process. The sinter yield is the ratio of the mass of + 5 mm sinter to all of the sinter. Tumbler index is a critical index to evaluate sinter strength, measured according to the criterion of ISO3271 (2007). Productivity is the finished sinter output per unit time and unit area. Solid fuel consumption is the mass of fuel consumption to produce finished sinter per unit mass.

Characterization

An optical microscope (LEICA MDI5000M, Germany) was used to observe the microstructure and mineral composition of sinter.

Results and Discussion

Effect of SiO_2 Content

Under the optimized conditions of moisture of 7.0%, suction negative pressure of 10 kPa and coke breeze dosage of 5.3%, SiO_2 content was investigated at basicity of 1.90 and return fines at proportion of 30% and the results are shown in Fig. 2. It was noted that SiO_2 content was decreased by reducing the proportion of E ore in blended ore, and the total proportion of coke breeze, return fines, fluxes, and blended

Fig. 2 Effect of SiO_2 content on sintering indexes. (Color figure online)

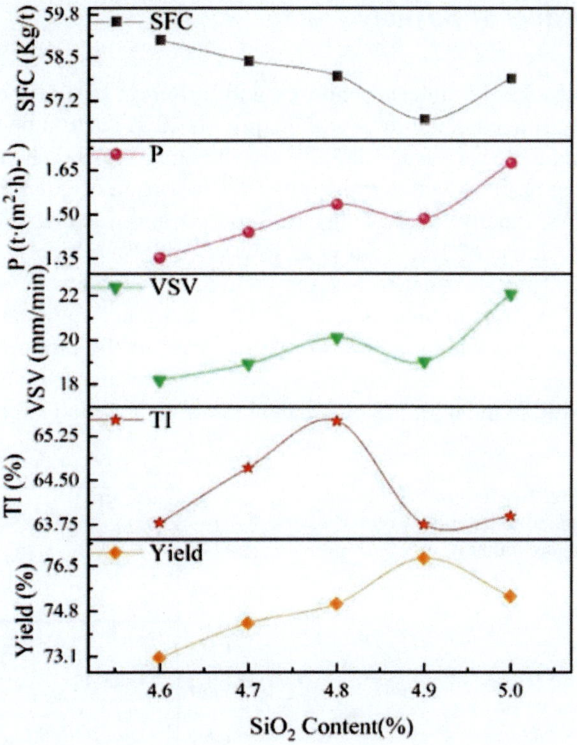

ore was 135.3%, which means coke breeze and return fines added in extra. Results show that with the decreasing of SiO_2 content, sintering indexes show a trend of getting worse. Productivity, VSV, TI, and yield decreased obviously, especially VSV decreased from 22.07 to 18.19 mm/min and yield decreased from 76.79 to 73.05%, which is the direct reason for the decrease of productivity. According to the structure of the raw material, it can be known that quicklime is the only way to adjust basicity in this research. To obtain a stable basicity , the proportion of quicklime has to be reduced when SiO_2 content decreased. However, quicklime is an important binder in the granulation process. As a result, the less quicklime was added in, the worse granulation of sinter mixture is, and the lower VSV is. The change tendency of solid fuel consumption with yield is often the opposite. In other words, when the yield increases, solid fuel consumption tends to decrease. At SiO_2 content of 4.9%, yield reached 76.79%, and SFC reduced to 57.95 kg/t. Tumbler index presents the tendency of going up firstly and going down secondly. That means at SiO_2 content of 4.8%, sinter could obtain good strength. Therefore, the lower limit of SiO_2 content was recommended at 4.8–4.9%.

Effect of Basicity

Results of sintering indexes with different basicity at SiO_2 content of 4.9%, and optimized conditions of moisture of 7.0%, suction negative pressure of 10 kPa and coke breeze dosage of 5.3% are shown in Fig. 3. When the SiO_2 content is constant, the basicity is determined by the proportion of quicklime. As shown in Fig. 3, with the basicity decreasing, the sintering indexes change mainly in a peak shape. At basicity of 1.8, yield is as low as 75.85%, which is close to others of 76.37 and 76.87%. In other words, the yield is at a high level. So, productivity increases with VSV, though yield decreases. TI shows a nearly linear decline with the decrease of basicity, which is because the decrease of the proportion of quicklime reduces the formation of calcium ferrite, an important binder phase in sintering. This indicates only in the high basicity can sinter obtain a good strength.

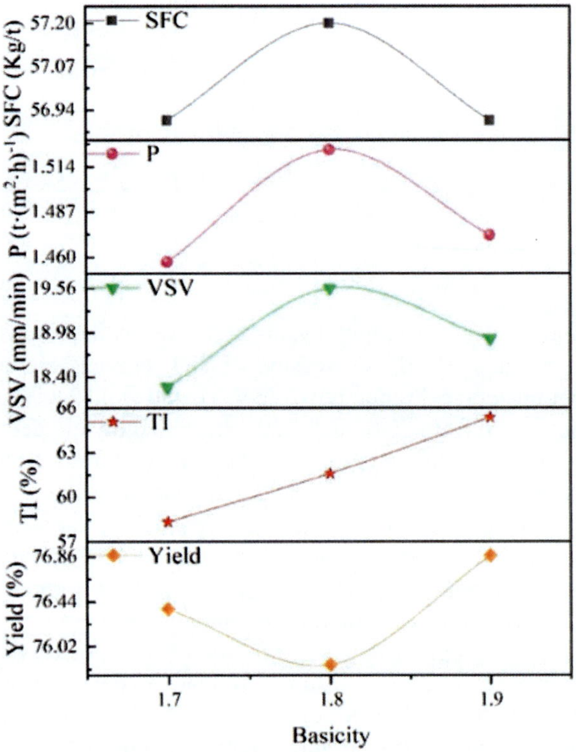

Fig. 3 Effect of basicity on sintering indexes. (Color figure online)

Comprehensive Influence of SiO_2 Content and Basicity

Under the condition of optimized conditions of moisture of 7.0%, suction negative pressure of 10 kPa and coke breeze dosage of 5.3%, effect of low SiO_2 content and low basicity on sintering indexes are shown in Table 2.

As shown in Table 2, at basicity of 1.7 and SiO_2 content of 4.60% tumbler index reaches the minimum value of 58.53%, which means at the low basicity and low SiO_2 content TI would further deterioration compared with single low SiO_2 content sintering or single low basicity sintering. At the same time, the results show that with the decrease of SiO_2 content, the sintering speed and the TI decreases and with the decrease of basicity, the sintering speed, the yield, and TI decrease.

The response surface of yield and TI at basicity from 1.70 to 1.90 and SiO_2 content from 4.58 to 4.98% are shown in Fig. 4, which gives a more intuitive change trend. TI is mainly affected by basicity for a significant decrease is shown in the graph. At high basicity, SiO_2 content has little effect on TI, while at low basicity TI changes in a saddle shape. Compared with TI, the range of yield change is much smaller, because it changes from 74.5 to 76.5%. Overall, yield decreases with the decline of SiO_2 content and basicity, and obtain the best value at basicity of 1.90 and SiO_2 content of 4.90%. This shows that low SiO_2 content and low basicity can be reached at a basicity of 1.80 and SiO_2 content of 4.60%.

Table 2 Sintering indexes of low SiO_2 content and low basicity sintering

SiO_2/%	Basicity	VSV mm/min	Yield/%	TI/%	P/t $(m^2\ h)^{-1}$	SFC/kg t^{-1}
4.70	1.70	20.66	74.05	60.03	1.593	59.32
4.60	1.70	18.56	75.41	58.53	1.472	57.91
4.70	1.80	20.59	75.46	61.03	1.610	58.18
4.60	1.80	20.24	76.07	60.09	1.605	57.70

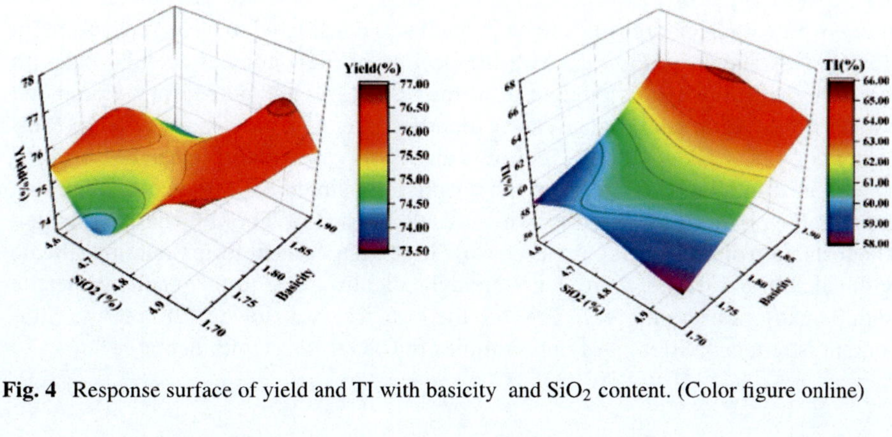

Fig. 4 Response surface of yield and TI with basicity and SiO_2 content. (Color figure online)

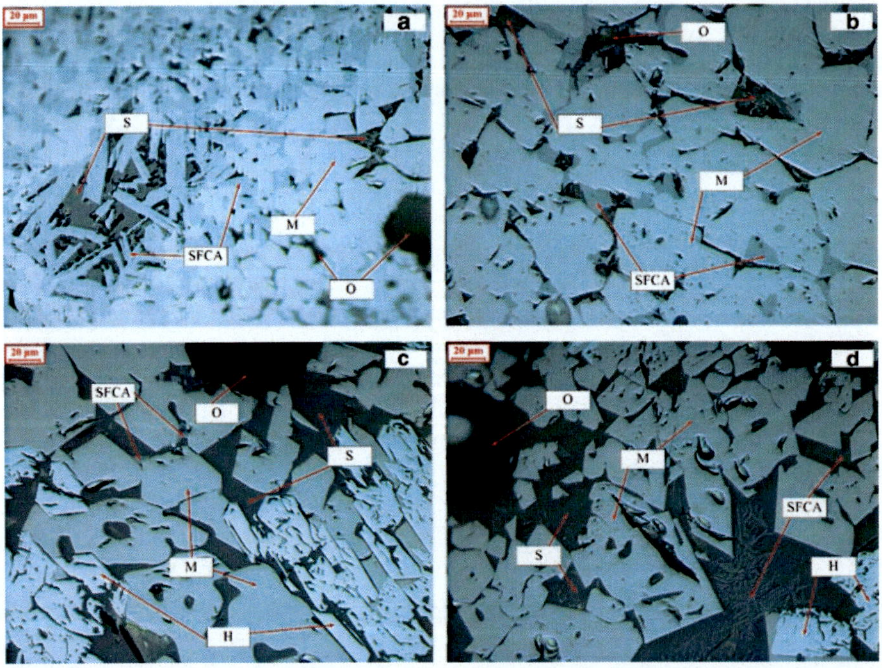

Fig. 5 The microstructures of sinters with different basicity and SiO_2 content. **a** basicity of 1.90, SiO_2 content of 4.98%; **b** basicity of 1.90, SiO_2 content of 4.68%; **c** basicity of 1.70, SiO_2 content of 4.90%; **d** basicity of 1.80, SiO_2 content of 4.60% (M, magnetite; H, hematite; S, silicate; O, pore). (Color figure online)

Analysis of Microstructure

The microstructure of sinter with different basicity and SiO_2 content is shown in Fig. 5. At basicity of 1.90 and SiO_2 content of 4.98%, acicular calcium ferrite was interwoven with hematite and silicate, which was considered to have good strength. However, as shown in Fig. 5b, with the decline of SiO_2 content acicular calcium ferrite decreased, which is thought of as the reason of yield decreasing. Compared with Fig. 5a, a significant change can be found in sinter with basicity of 1.70 and SiO_2 content of 4.90% shown in Fig. 5c. It was almost no calcium ferrite can be observed in Fig. 5c, and silicate is the main binder phase in sinter. Silicate is considered an unfavorable phase for its fragility, which results in the low TI of low basicity sinter. The formation of such a structure is because high SiO_2 content forms calcium silicate with calcium oxide, and there is not enough calcium oxide to form calcium ferrite with hematite. Furthermore, in Fig. 5d, the basicity is as low as 1.8, but the SiO_2 content is reduced. So it gives opportunities to CaO to react with hematite.

Conclusion

Through the systematic study of low SiO_2 content and low basicity sintering, it can be found that with the decline of basicity, TI decreases obviously while the yield shows different rules that at high SiO_2 content yield can obtain good value at low basicity and high basicity, and at low SiO_2 content yield can get good result at basicity of 1.80. Furthermore, the effect of SiO_2 content on sinter quality is not as significant as basicity, and when SiO_2 content ranges from 4.70 to 4.80%, then sinter could obtain a qualified tumbler index. The rule of low SiO_2 content and low basicity provides a direction of further research and gives a guidance to the implementation of low SiO_2 content and low basicity sintering.

Acknowledgements This work is supported by the National Key Research and Development Program of China (2017YFB0304301).

References

1. Lu L, Ishiyama O (2016) Recent advances in iron ore sintering. Miner Process Extr Metall 125:132–139
2. World Steel Association (2020) 2020 world steel in figures. https://www.worldsteel.org/en/dam/jcr:f7982217-cfde-4fdc-8ba0-795ed807f513/World%2520Steel%2520in%2520Figures%25202020i.pdf. Accessed 1 Sept 2020
3. Ma D, Chen W, Yin X (2016) Quantifying the co-benefits of decarbonization in China's steel sector: an integrated assessment approach. Appl Energ 162:1225–1237
4. Guo ZC, Fu ZX (2010) Current situation of energy consumption and measures taken for energy saving in the iron and steel industry in China. Energy 35(11):4356–4360
5. Jiang T (ed) (2015) Principle and technology of agglomeration of iron ores. Central South University Press, Changsha
6. Huang XG (ed) (2013) Principles of iron and steel metallurgy. Metallurgical Industry Press, Beijing
7. Feng X, Zhang Y, Li Z (2004) Influence of basicity on the low silicon sinter strength. Sinter Pelletizing 29(2):9–11
8. Wu Y, Li X, Yu Z (2018) Study on metallurgical performance and optimized blending of low silicon sinter. J Anhui Univ Metal Technol 28:19–22
9. Zhou Y, Kang X, Liu G (2010) Study on high Fe and low SiO_2 sintering technology. Sinter Pelletizing 35(3):20–24

Part IV
Material Processing and Recycling

Recovery of Zinc from Oxide-Sulphide Zinc Ore Through Oxidation and Chelation

Kun Yang, Chengyu Sun, Hongtao Qu, Likun Shuo, Yongguang Luo, Libo Zhang, and Aiyuan Ma

Abstract In this paper, three different processes—chelation accompanied by oxidation, oxidation followed by chelation, and oxidative acid leaching followed by chelation—are designed for extraction of zinc from oxide-sulphide zinc ore, and through a comparison of the redox potentials, suitable oxidation additives are determined for each process. Besides, effects of oxidant additives doping level on zinc recovery are investigated. The results show that the zinc leaching degree can reach 83.72% with addition of NaO being 0.3 mol/L in the process of chelation accompanied by oxidation, the zinc leaching degree can reach 92.40% with NaO adding 0.5 mol/L in the process of oxidation followed by chelation, and the zinc leaching degree can reach 93.02% under a condition of NaClO doping level being 0.1 mol/L in the process of oxidative acid leaching followed by chelation. All three processes prove their value on extraction of oxide-sulphide zinc ores.

Keywords Chelation · Oxidation · Oxidative acid leaching · Extractive · Oxide-sulphide zinc ore

---

K. Yang · C. Sun (✉) · H. Qu · L. Shuo · Y. Luo · L. Zhang
Faculty of Metallurgical and Energy Engineering, Kunming University of Science and Technology, Kunming 650093, Yunnan, China
e-mail: qjlyg@163.com

C. Sun · H. Qu · L. Shuo · Y. Luo
Yunnan Chihong Zn&Ge Co., LTD, Qujing 655011, Yunnan, China

K. Yang · L. Zhang
National Local Joint Laboratory of Engineering Application of Microwave Energy and Equipment Technology, Kunming 650093, Yunnan, China

A. Ma
School of Chemistry and Materials Engineering, Liupanshui Normal University, Liupanshui 553004, China

B. Li et al. (eds.), *Materials Engineering—From Ideas to Practice: An EPD Symposium in Honor of Jiann-Yang Hwang*, The Minerals, Metals & Materials Series,
https://doi.org/10.1007/978-3-030-65241-8_15

Introduction

Zinc is an important strategic nonferrous metal resource. With the rapid growth of economic, there is a gradually increasing demand for zinc, while finite reserved sphalerite-traditional zinc resource is unable to meet the demand [1]. Now the non-traditional zinc resources like zinc oxide ore, oxide-sulphide zinc ore and zinc metallurgical slag and dust have drawn the attention of scientists, and have been recognized as valuable processing nonferrous resources [2].

Zhang et al. [3] had studied recovering zinc from solid waste bearing sphalerite or zinc ferrite by mechano-chemical extraction in alkaline solution and got a conclusion that mechano-chemical leaching is more efficient in comparison with the chemical leaching of non-activated sample. Mehrabani et al. [4] investigated the leaching of lead–zinc tailing dam and stated that sphalerite was successfully dissolved in preference to the pyrite using moderate thermophile bacteria in lower redox potential. Xia and Pickles [5] observed that the leaching kinetics of zinc and lead in caustic solution were significantly improved when microwave energy was used as a heating source. Ma et al. [6] adopted a new process of alkaline leaching-electrowinning to recovery low-grade zinc oxide ore and realized the separation of zinc and impurities. Zhang et al. [7] proposed ammonium sulphate technology for the complex zinc dust with high F and Cl.

Although there are many reports on the recovery of zinc oxide ore or zinc metallurgical slag and dust [3, 8, 9], researches on extractive of oxide-sulphide zinc ore are scarce. On this basis, three different processes are designed, which are enhanced by adding oxidant, and appropriate doping levels of oxidations are investigated for each process. What's more, oxidizing mechanisms are also briefly discussed. Compared with the traditional process of recovery of valuable metal zinc from oxy-sulphur mixed ore, the economic cost is high. The aim of this paper is to find a cost-effective and eco-friendly process for recovery of oxide-sulphide zinc ore.

Experimental

Materials

The oxide-sulphide zinc ore used for this study is from Lanping county of China. Its main chemical compositions and zinc phase distributions are given in Tables 1 and

Table 1 Main chemical composition of complex zinc ore (mass fraction, %)

Zn_T	CaO	SiO_2	S_T	Fe_T	MgO	Pb	Al_2O_3
6.45	28.74	12.54	1.43	3.91	0.45	4.31	1.57

Zn_T total zinc; S_T total sulphur; Fe_T total Fe

Table 2 Zinc phase distributions of oxide-sulphide zinc ore

Zn	ZnO	ZnS	$ZnFe_2O_4$ etc	$ZnSO_4$	Zn_T
Mass fraction/%	5.68	0.53	0.21	0.03	6.45
Distribution/%	88.06	8.22	3.26	0.46	100%

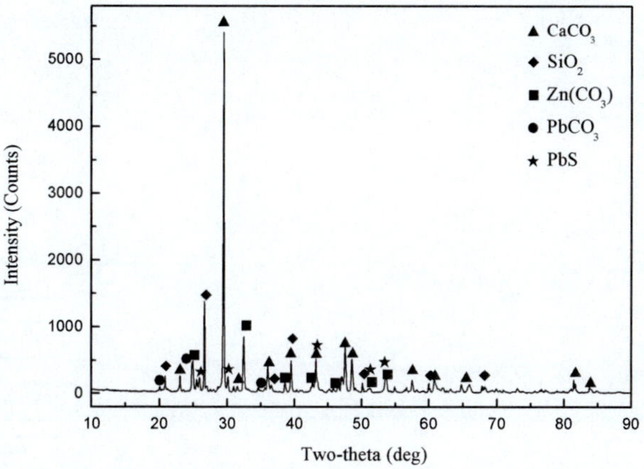

Fig. 1 XRD pattern of oxide-sulphide zinc ore

2, respectively, which demonstrate that this zinc ore has a low grade, while its ratio of zinc oxides/zinc sulphide is high.

XRD technique was applied to analyse the phase composition, and the result is shown in Fig. 1. From Fig. 1, it can be concluded that main mineralogical phases in this oxide-sulphide zinc ore include calcium carbonate [$CaCO_3$], zinc carbonate [$ZnCO_3$], cerussite [$PbCO_3$], silicon dioxide [SiO_2], and galena [PbS]. Distilled water and reagent grade chemicals were used to prepare all the leaching solutions.

Leaching Process

The leaching experiments are carried out in a CJJ-931/HJ-6 six-connected-magnetic stirrers with a speed range of 0–2400 rpm by the way of mechanical agitation, which are all happened in covered Pyrex Erlenmeyer flasks. Based on the sequence of oxidation and chelation, three different technologies were designed, as shown in Fig. 2a–c.

From Fig. 2, it can be seen that all these three processes consist of two stages, raw ore leaching and leached residual leaching. At the first stage, raw material was leached by the filtrate solution produced in the second stage within one hour. And at the second stage, the filter residual is leached by the prepared ammonium chloride solution ($c(NH_3)_T = 7.5$ mol/L, $c(NH_4Cl):c(NH_4OH) = 1:1$), and the filtrate will be

Fig. 2 Different technological processes for recovering complex zinc ore

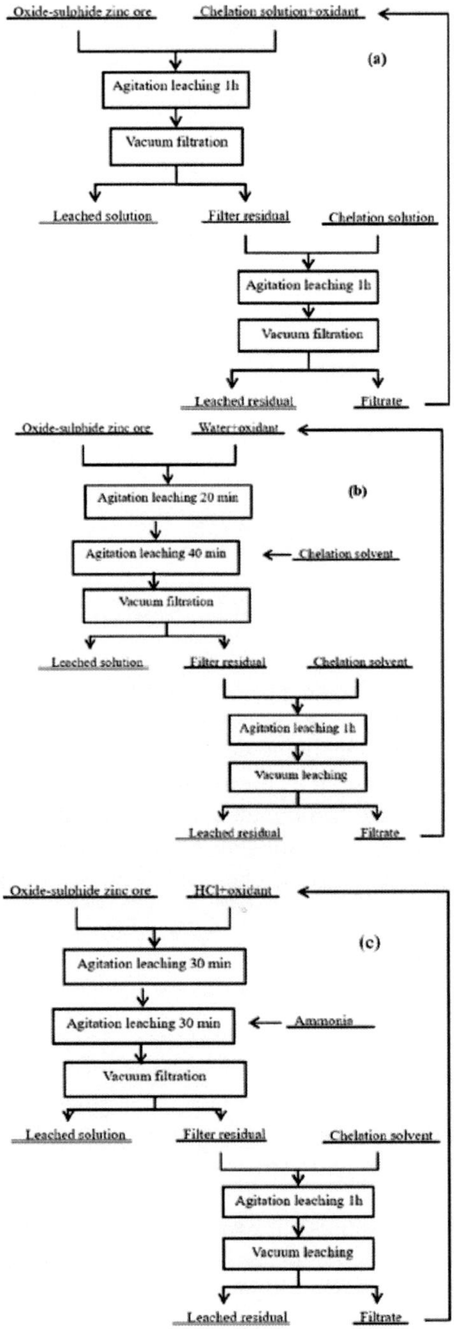

Table 3 Property of the oxidants

Species	H_2O_2	$KMnO_4$	Na_2O_2
Property	Acid medium oxidant	Acid medium oxidant	Alkali medium oxidant
Species	NaClO	$CuCl_2{\cdot}2H_2O$	$FeCl_3{\cdot}6H_2O$
Property	Alkali medium oxidant	Acid medium oxidant	Acid medium oxidant

recycled to apply in the first stage. All three processes are developed for circulating leaching.

The essential distinctions between three procedures exist in the first stage, process (a) characterizing as chelation accompanied by oxidation, namely, adding oxidant and chelated solution simultaneously [10, 11], process (b) featuring as oxidation followed by chelation—oxidizing zinc ore 20 min, and then adding chelation solvents to leach 40 min, and the trait of process (c) being oxidative acid leaching followed by chelation, in other words adding HCl and oxidant oxidizing 30 min first [12–15], and injecting ammonia to chelated 30 min followed.

Agitation was at the rate of 300 rpm, leaching temperature was 303 K, the ratio of solid and liquid was 10:1, and leaching solution was ammonium chloride. Total concentration of ammonium was 7.5 mol/L, and the molar ratio of NH_4Cl and $NH_3{\cdot}H_2O$ was 1:1. Meanwhile, oxidant is added to improve zinc recovery rate, adding amounts of oxidant choosing 0, 0.1, 0.3, 0.5, 0.8, 1 mol/L, respectively.

Oxidizing Mechanism

The oxidants explored in this research include H_2O_2, K_2MnO_7, NaClO, Na_2O_2, $FeCl_3{\cdot}6H_2O$ and $CuCl_2{\cdot}2H_2O$, and their properties display at Table 3.

Prepare the leaching solution in a molar ratio of NH_4Cl and $NH_3{\cdot}H_2O$ being 1:1, and pH value of this solution can be defined as

$$pH = pKa = 14 - pKb \tag{1}$$

Thus, theoretical pH value of leaching solution is calculated:

$$pH = 14 - pKb = 14 - \left[-\log\left(1.8 * 10^{-5}\right)\right] = 9.26 \tag{2}$$

What's more, the measured pH value of ammonium chloride solution is about 9.3. To simplify calculation, the pH value of ammonium chloride solution is taken as 9.3 regardless of the oxidation doping impact.

Besides, potentials of oxidations in alkaline solution and acid solution are, respectively, referred as.

H_2O_2—In alkaline solution: $H_2O_2 + 2e = 2OH^- E = 0.88$ V;

In acid solution: $H_2O_2 + 2H^+ + 2e = 2H_2O E = 1.77\ V$;

$KMnO_4$—In alkaline solution: $MnO_4^- + 2H_2O + 3e = MnO_2 + 4OH^- E = 0.588$ V;

In acid solution: $MnO_4^- + 4H^+ + 3e = MnO_2(s) + 2H_2O E = 1.695$ V;

NaClO—in alkaline solution: $ClO^- + H_2O + 2e = Cl^- + 2OH^- E = 0.89$ V;

In acid solution: $HClO^- + H^+ + e = 1/2\ Cl_2 + H_2O^- E = 1.63$ V;

$CuCl_2 \cdot 2H_2O - Cu^{2+} + e = Cu^- E = 0.519$ V (The pH value for Cu^{2+} totally precipitation is 6.4);

$FeCl_3 \cdot 6H_2O - Fe^{3+} + e = Fe^{2+} E = 0.771$ V (The pH value for Fe^{3+} totally precipitation is 3.2);

Meanwhile the reducing potential of S is

$$S - S^{2-} - 2e = SE = 0.48\ V$$

Based on the above data, H_2O_2, K_2MnO_7, Na_2O_2, and NaClO are suitable for all three processes, while $CuCl_2 \cdot 2H_2O$ and $FeCl_3 \cdot 6H_2O$ just can be applied in process (c).

Results and Discussion

Leaching rates of zinc in the process (a) are illustrated in Fig. 3a. From it, conclusion can be drawn as oxidants—H_2O_2 and Na_2O_2 can somewhat improve the recovery of oxide-sulphide zinc ore, oxidants–NaClO and $KMnO_4$ show no effect on ammonia leaching of this ore, while oxidants–$CuCl_2 \cdot H_2O$ and $FeCl_3 \cdot 6H_2O$ will lower its leaching. When leach with doping H2O2 or Na_2O_2, the leaching rates show a trend of increasing firstly, then dropping partly, and zinc leaching rate can reach 83.72% with Na_2O_2 adding content being 0.3 mol/L. In conclusion, process (a) is feasible to treat this zinc ore, and recovery of zinc reaches above 80% on condition of N_2O_2 adding amount being 0.3 mol/L.

Figure 3b indicates the effects of various oxidations on extractive of oxide-sulphide zinc ore in the process (b). Judged from the curves' trend, only Na_2O_2 is beneficial for this zinc ore, and NaClO and H_2O_2 offer no value. With Na_2O_2 adding, leaching rate of zinc gradually improves, while growth trends become slow, when adding amount exceeds 0.5 mol/L, and maxim leaching rate obtained is 92.40% at this point. Other oxidations, such as K_2MnO_7, $FeCl_3 \cdot 6H_2O$ or $CuCl_2 \cdot 2H_2O$ are found causing serious deposition phenomena, which will present a big problem for the vacuum leaching.

The behaviours of oxidants in the process (c) are also investigated, and results are shown in Fig. 3c. Through a comprehensive comparison, it can be concluded that H_2O_2 or NaClO belongs to suitable oxidations for this process, leaching rates of oxide-sulphide zinc ore get obvious improvement, and when the adding amount of NaClO (H_2O_2) is 0.1 mol/L, leaching rate is 93.02% (84.50%). While other oxidations will lower the leaching or can't be applied in this process.

Fig. 3 Recovery rates of oxide-sulphide zinc ore in three processes. (Color figure online)

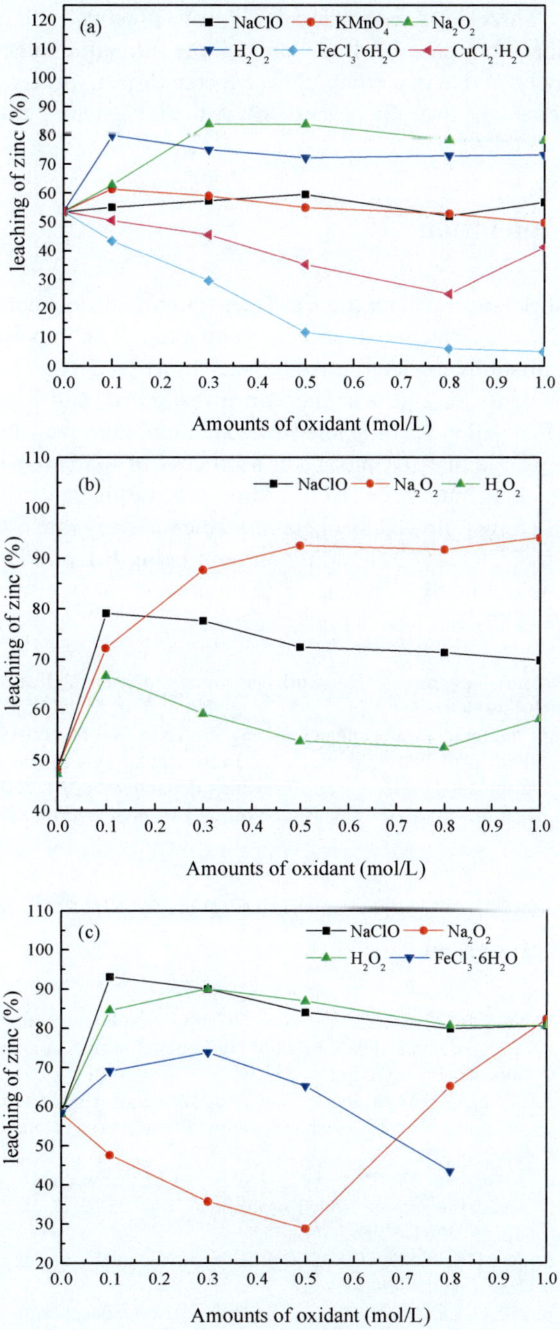

Three designed processes all prove practical, and under suitable oxidation conditions, they can effectively prevent the formation of gel-SiO_2, thus have a high selectivity. While in a comprehensive comparison, process (b) and (c) shows an obvious advantage over the process (a), with zinc leaching rate beyond 90%.

Conclusion

This paper explores three different circulation technologies to recover oxide-sulphide zinc ore through oxidation and chelation. By comparing the redox potential, suitable oxidations are determined for each process. On this basis, effects of various common oxidants on zinc leaching are investigated, and it can be found that in a process of chelation accompanied by oxidation, zinc recovery rate can reach 83.72% with N_2O_2 adding 0.3 mol/L, in a process of oxidation followed by chelation, perfect leaching rate with N_2O_2 adding 0.5 mol/L, and in the process of oxidative acid leaching followed by chelation, zinc recovery rate is 93.02% (84.50%) on condition of NaClO (H2O2) adding amount being 0.1 mol/L. Though the optimal leaching rate is slightly different, all three processes can effectively prevent the formation of gel-SiO_2 and have a high selectivity.

Acknowledgements This work was supported by Discipline construction funds for Introduction of high-level talents (130214119512), the Nonferrous Metal Electrodeposition Technology Provincial Innovation Team of Yunnan Chihong Zn & Ge Co., Ltd (201905E160007), and Cina Scholarship council (CSC 202008530041), the Liupanshui Key Laboratory of Metallurgical Energy Saving, Environmental Protection and Recycling Economy (52020-2018-0304), the Science and Technology Innovation Group of Liupanshui Normal University (LPSSYKJTD201801).

References

1. Peng R (2005) Zinc metallurgy. Central South University Press, Changsha
2. Yang D, Zhu H, Chen J (2006) Hydrometallurgical zinc process and technology. Metallurgical Industry Press, Beijing
3. Zhang C, Wang J, Bai J, Zhao Y (2012) Recovering of zinc from solid waste bearing sphalerite or zinc ferrite by mechano-chemical extraction in alkaline solution. Procedia Environ Sci 16:786–790
4. Mehrabani JV, Shafaei SZ, Noaparast M, Mousavi SM, Rajaei MM (2013) Bioleaching of sphalerite sample from Kooshk lead–zinc tailing dam. Trans Nonferrous Metals Soc China 12(23):3763–3769
5. Xia DK, Pickles CA (2000) Microwave caustic leaching of electric arc furnace dust. Miner Eng 13(1):79–94
6. Ma Q, Li X, Chen S, Li X, Zhang F (2012) One technology to process the low-grade zinc oxide ores. China patent C02133784.5. 17 September 2002
7. Zhang M, Zhang P, He J, Yuan X, Chen Y (1998) Leaching zinc dust in system of Zn(II)–$(NH_4)_2SO_4$–H_2O. J Central South Univ J Central South Univ 8(1):118–121

8. Ma AY, Zheng XM, Li S, Wang YH, Zhu S (2018) Zinc recovery from metallurgical slag and dust by coordination leaching in NH_3–CH_3COONH_4–H_2O system. Roy Soc Open Sci 5(7):180660
9. Ma AY, Zheng XM, Zhang LB, Peng JH, Li Z, Li S, Li SW (2018) Clean recycling of zinc from blast furnace dust with ammonium acetate as complexing agents. Sep Sci Technol 53(9):1327–1341
10. Ghosh MK, Das RP, Biswas AK (2002) Oxidative ammonia leaching of sphalerite part i: noncatalytic kinetics. Int J Miner Process 66:241–254
11. Ghosh MK, Das RP, Biswas AK (2003) Oxidative ammonia leaching of sphalerite part II: Cu(II)-catalyzed kinetics. Int J Miner Process 70:221–234
12. Uçar G (2009) Kinetics of sphalerite dissolution by sodium chlorate in hydrochloric acid. Hydrometallurgy 95:39–43
13. Dehghan R, Noaparast M, Kolahdoozan M (2009) Leaching and kinetic modelling of low-grade calcareous sphalerite in acidic ferric chloride solution. Hydrometallurgy 96:275–282
14. Aydogan S, Aras A, Canbazoglu M (2005) Dissolution kinetics of sphalerite in acidic ferric chloride leaching. Chem Eng J 114:67–72
15. Aydogan S (2009) Dissolution kinetics of sphalerite with hydrogen peroxide in sulphuric acid medium. Chem Eng J 123:65–70

Research and Industrial Application of the Evaluation Method of Pulverized Coal Injection for Blast Furnace

Dongqing Wang, Weichun Zhu, Yapeng Zhang, and Jinhua Wang

Abstract The evaluation method for pulverized coal injection in blast furnace including parameters of blast volume, pressure difference, gas dust production, quality, etc. was developed in this work. Indexes for smooth operation of blast furnace and utilization of PCI (pulverized coal injection) coal (named carbon content index) were also proposed. The evaluation method can reflect the effect of type of pulverized coal on the utilization ratio of pulverized coal and the blast furnace performances. The industrial trial showed that when 10% Coal B was used, smooth operation index reduced by more than 0.05 and carbon content index increased by over 1.0, which indicated that coal B could not be used too much in PCI. When coal A was used up to 35%, smooth operation index increased 0.08 and carbon content index reduced by over 0.22 with blast furnace output and coal ratio increasing, which meant that Coal A was suitable for PCI.

Keywords Blast furnace · Pulverized coal injection · Utilization · Evaluation method

Introduction

The total output of blast furnace ironmaking in China was 771.05 million tons in 2018. According to the coal ratio of 139.4 kg/t (statistical data of 344 blast furnaces), 107.48 million tons of raw coal were needed. In recent years, the price of coke has repeatedly reached new heights and the pressure of using pulverized coal injection (PCI) to partially replace the coke to reduce costs is increasing. As a result, the supply of pulverized coal has become tighter and tighter, and it is necessary to develop new resources of pulverized coal for PCI. Laboratory studies on new pulverized coal

D. Wang (✉) · W. Zhu · Y. Zhang · J. Wang
Beijing Key Laboratory of Green Recyclable Process for Iron & Steel Production Technology, Beijing 100043, China
e-mail: wangdongqing1980@163.com

Shougang Research Institute of Technology, Beijing 100043, China

B. Li et al. (eds.), *Materials Engineering—From Ideas to Practice: An EPD Symposium in Honor of Jiann-Yang Hwang*, The Minerals, Metals & Materials Series,
https://doi.org/10.1007/978-3-030-65241-8_16

resources mainly focus on the properties of single coal, such as proximate analysis, ultimate analysis, ash composition analysis, combustion property, and grindability [1–3]. Only the utilization of pulverized coal was analyzed, such as utilization ratio and replacement ratio of pulverized coal [4–6]. But the influence of PCI on the smooth operation of blast furnace was rarely reported. In this paper, an evaluation method for pulverized coal used in industrial tests is proposed. This method can reflect the influence of pulverized coal switch on utilization ratio of pulverized coal and blast furnace performance.

The Evaluation Method for Pulverized Coal

The evaluation method of pulverized coal for the industrial test is mainly about the evaluations of blast furnace performance and pulverized coal utilization under high coal ratio and high utilization coefficient of blast furnace, which includes the blast furnace smooth operation index and carbon content index.

Blast Furnace Smooth Operation Index

$$ID_s = k_1 * V / \Delta P * (P * PCI) \tag{1}$$

ID_s—Smooth operation index; V—Blast volume, m^3/min; ΔP—Total pressure difference, kPa; P—Daily output, t/d; PCI—Coal ratio, kg/t; k_1—Coefficient 1.

In Eq. 1, factors such as blast volume, total pressure difference, daily output, and coal ratio are taken into account in the calculation of the smooth operation index. Blast volume and daily output represent the smelting capacity and smelting performance of blast furnace, respectively. The total pressure difference represents the permeability of blast furnace on the premise of a certain blast volume. And coal ratio represents the coke-saving level of blast furnace. These parameters are interrelated. The larger the smooth operation index is, the better the blast furnace operation state is, and more technical and economic performances will be obtained. Under the conditions of stable raw materials, fuels and basic operations, the effects of type switch of pulverized coal on the utilization ratio can reflect the influence of a new coal on blast furnace performance.

Carbon Content Index

$$C_w = P_x * C_x + P_g * C_g \tag{2}$$

C_w—The total carbon content in gas dust, kg/t; P_x—Gravitational dust ratio, kg/t; C_x—Carbon content in gravitational dust, %; P_g—Bag dust ratio, kg/t; C_g—Carbon content in bag dust, %;

$$ID_t = k_2 * C_w / (PCI * V) \quad (3)$$

ID_t—Carbon content index; k_2—Coefficient 2.

In Eq. 2, C_w is the sum of gravitational dust carbon and bag dust carbon. In Eq. 3, the carbon content index takes into account total carbon content, coal ratio, and blast volume. If the quality of coke and raw materials are basically stable, the total carbon content in gas dust reflects the quality of unburned pulverized coal. Carbon content index takes into account the effect of coal ratio and blast volume on the total carbon content. According to the change of carbon content index in different periods, the effects of type switch of pulverized coal on unburned coal can be evaluated.

The Application of the Evaluation Method

Laboratory Investigation

The results of proximate analysis are essential in PCI process. The Chinese Standard GB/T 212 was used in the proximate analysis of pulverized coals. The results of the proximate analysis are shown in Table 1. Coal A and B are new coals used in the tests. The other coals are currently used in Shougang. Table 2 shows elemental analysis of the five coals.

Table 1 Proximate analysis (dried)

Coal	FC_d %	V_d %	A_d %	$Q_{b.d}$ Mj/kg	S_{td}, %
Yang	76.25	11.33	12.42	31.95	0.52
Shen	57.62	33.60	8.78	27.09	0.18
Lu	77.87	11.25	10.88	31.63	0.37
A	79.15	11.07	9.78	31.88	0.51
B	83.94	3.41	12.65	28.57	0.41

Table 2 Ultimate analysis (%)

Coal	C_d	H_d	O_d	N_d
Yang	80.98	3.52	1.50	1.06
Shen	72.39	4.08	13.32	1.25
Lu	80.16	3.42	1.63	1.19
A	81.78	3.41	2.82	1.70
B	83.04	0.93	2.57	0.40

It shows that coal A and B were anthracites according to Tables 1 and 2. The fixed carbon in coal A is higher than coal Yang. Ash content of coal A is lower than coal Yang, while calorific value and sulfur content are close to the Yang coal. Hydrogen content in coal A is lower than coal Yang. Coal B has higher F_{Cd} and A_d, also lower V_d and H_d.

Coal ash composition has a great influence on BF slag liquidity and desulfurizing at a large PCI rate. In Table 3, it can be seen that ash components of coal A and coal B are similar, showing acid and Al_2O_3 contents of the two coals are lower than the other coals. Alkalis contents in coals A and B are higher (coal B has a higher K_2O content and coal A has a higher Na_2O content).

The liquidus temperature of the five coal ashes are calculated by FactSage 7.1 and shown in Fig. 1. It shows that the liquidus temperatures of ashes in coal A and coal B are higher than that of coal Shen ash, but lower than those of ashes in coals of Yang and Lu. It is speculated that the ashes of coal A and coal B are not easily to melt compared with that of coal Shen at the tip of tuyere.

Grindability represents whether the coal is easy to grind or not. The test of coal grindability followed Chinese Standard GB/T2565-1998. Table 4 shows that both coal A and coal B have high grindability.

Table 3 Pulverized coal ash composition analysis %

Coal	SiO_2	Al_2O_3	Fe_2O_3	CaO	MgO	K_2O	Na_2O
Yang	52.97	39.83	1.74	1.48	0.58	0.38	0.60
Shen	38.48	30.86	1.21	17.9	2.22	0.09	2.51
Lu	42.10	38.40	4.32	4.89	0.47	1.37	0.76
A	54.91	29.03	3.31	3.36	1.69	0.71	2.17
B	57.40	29.55	4.23	1.08	1.23	2.92	0.93

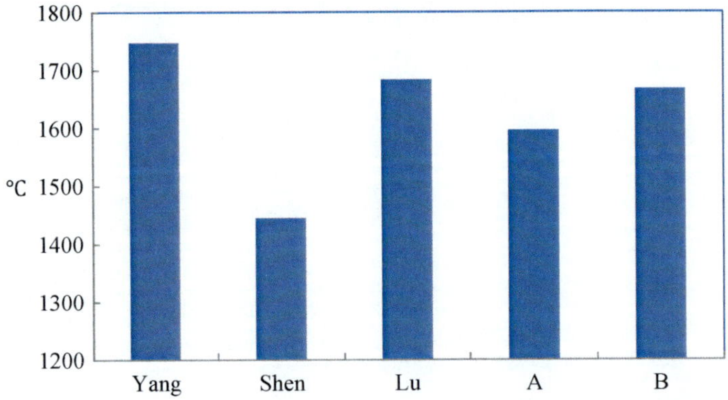

Fig. 1 The liquidus temperature of the five coal ash. (Color figure online)

Table 4 Pulverized coal grindability index

Coal	Yang	Shen	Lu	A	B
Grindability	75	80	74	90	116

Table 5 Ignition point results

Coal	Yang	Shen	Lu	A	B
Ignition point, °C	490	331	514	465	564

Table 5 shows that the ignition point of coal A is 465 °C, which is medium compared with others. The ignition point of coal B is 564 °C which shows a difficult firing ability.

Table 6 shows differential thermal analysis of the coal samples. The comprehensive combustion characteristic index of coal A is close to coal Lu. The comprehensive combustion characteristic index of coal B is small, which reflecting bad combustion performance.

Table 7 shows the theoretical replacement ratios of coals. The theoretical replacement ratio is calculated according to Shougang blast furnace condition. The theoretical replacement ratio of coal A is 0.95, which is higher than other coals. The theoretical replacement ratio of coal B is 0.85, which is lower than other coals.

The above analysis shows coal A has higher fixed carbon content, lower ash content, better grindability, and higher theoretical replacement ratio than other coals. That means coal A is suitable for PCI. Coal B has a higher fixed carbon content and ash content, lower volatilization and calorific value, worse combustion than other coals. So coal B is not very good for PCI.

Table 6 Analysis of combustion feature

	Ignition point °C	Maximal combustion point °C	Burn-out temperature °C	Maximal combustion rate %/min	Average combustion rate %/min	Comprehensive combustion characteristic index 10^{-9} mg^2 min^{-2} °C^{-3}
Yang	490	629	893	5.76	4.15	1.12
Shen	345	477	739	5.97	4.69	3.19
Lu	514	641	977	5.11	3.56	0.74
A	465	628	973	4.80	3.36	0.77
B	564	752	1197	2.59	2.29	0.16

Table 7 Theoretical replacement ratio kg/kg

Coal	Yang	Shen	Lu	A	B
Theoretical replacement ratio	0.93	0.87	0.93	0.95	0.85

Industrial Trial

Industrial Trial of Coal A

During Trial period 1 in the trial, the proportion of coal A was 22%. The proportion of coal A reached 35% in Stage 2 of the trial. Coal A replaced coal Yang.

The technical and economical indexes of blast furnace are shown in Table 8. It can be seen materials and fuels were stable. During the trial period, output and coal ratio of blast furnace raised, and the coke ratio reduced. Furthermore, blast volume raised over 50 m^3/min, and gas utilization rate reduced over 0.7%.

The evaluation method for pulverized coal was applied to analyze whether coal A was suitable for PCI process. According to the evaluation method, the blast furnace smooth operation index (ID_s) raised by 0.08, the total carbon content in gas dust (C_w) reduced by 0.22. It showed the blast furnace performance and coal utilization is better when Coal A increased to 35%. Hence, coal A is suitable for injection in blast furnace.

Table 8 Technical and economical indexes of blast furnace in Shougang

Item	Benchmark	Trial period 1	Trial period 2
Blast volume m^3/min	8348	8410	8428
Blast temperature °C	1241	1246	1247
Permeability index	3952	3906	3967
Output t/d	13,041	13,134	13,197
Hot metal temperature °C	1500	1520	1510
Gas utilization %	50.2	49.44	49.06
Sinter drum index %	82.73	82.67	82.71
Pellet compressive strength N	2612	2625	2659
Coke ash %	11.99	12.1	11.97
M40 %	90.77	90.88	91.03
Pulverized coal ash %	10.12	10.16	9.83
Pulverized coal V_d %	17.34	16.68	16.25
Fuel ratio kg/t	498.2	496.9	491.3
Coke ratio kg/t	297.7	275.4	271.5
Coal ratio kg/t	179.8	197.4	196.5
C_W	1.74	1.8	1.51
ID_t	3.48	3.26	2.73
ID_s	0.93	1.01	1.03

Table 9 Technical and economical index of blast furnace in Shougang

Item	Benchmark	Trial period
Blast volume (m^3/min)	8418	8319
Blast temperature (°C)	1248	1247
Permeability index	4275	4161
Output (t/d)	13,188	12,863
Hot metal temperature (°C)	1501	1517
Gas utilization (%)	50	47.9
Sinter drum index (%)	82.59	82.55
Pellet compressive strength (N)	2653	2616
Coke ash (%)	11.82	11.85
M_{40} (%)	92.15	92.11
Pulverized coal ash (%)	9.4	9.41
Pulverized coal V_d (%)	20.15	19.58
Fuel ratio (kg/t)	485.1	480.4
Coke ratio (kg/t)	271	269.6
Coal ratio (kg/t)	195.6	192.4
C_W	1.27	1.79
ID_t	2.32	3.35
ID_s	1.1	1.03

Industrial Trial of Coal B

An industrial trial was carried out in a blast furnace of Shougang. During the trial period, the proportion of coal B was 10%, replacing coal Yang. Considering to reduce the fluctuation of blast furnace conditions, benchmark was chosen when raw material and coke quality was relatively stable.

The materials quality and technical indicators of the blast furnace are shown in Table 9. It can be seen that coke ash, pellet strength, sinter strength were stable. In the trial period, the blast furnace output and coal ratio reduced a little. Blast volume reduced by 100 m^3/min and gas utilization rate reduced by 1.0%.

The evaluation method for pulverized coal was applied to analyze whether coal B was suitable for PCI process. Compared with benchmark, the blast furnace smooth operation index (ID_s) reduced by over 0.05. The total carbon content in gas dust (C_w) raised by over 0.5, and the carbon content index (ID_t) raised by over 1.0. It showed that blast furnace performance and coal utilization got worse when the ratio of coal B went up to 10%. As a result, Coal B was not recommended to be used in PCI.

Analysis

When coal B was injected into blast furnace, the permeability index of BF decreased, which is related to the high ignition point and low combustion characteristic index of coal B. Although the proportion of coal B injection was only 10%, a lot of unburned pulverized coal formed in the blast furnace. The permeability of the blast furnace was seriously affected, and resulted in the decrease of the blast furnace smooth operation index (ID_s) and the increase of the carbon content index (ID_t).

When coal B was injected into blast furnace, the permeability index of blast furnace maintained stable on the premise of higher output and coal ratio, which was related to the lower ignition point and lower combustion characteristic index of coal A. Although the proportion of injected Coal A was up to 35%, the total carbon content (C_w) and carbon content index (ID_t) were reduced. It shows that with the increase of output and coal ratio, the unburned pulverized coal in the furnace did not increase, the permeability of the blast furnace remained stable, and the performance of the blast furnace improved.

Conclusions

(1) The blast furnace pulverized coal evaluation method was developed in this work. This method can be used to reflect the influence of type switch of pulverized coal on the utilization of coal and the blast furnace performance. The smooth operation index (ID_s) and carbon content index were proposed. The method takes into account blast furnace operation parameters (blast volume, pressure difference, etc.), gas dust, etc. The blast furnace performance and utilization ratio of pulverized coal were evaluated effectively.
(2) When 10% coal B was used, smooth operation index reduced (ID_s) by over 0.05 and carbon content index (ID_t) increased by over 1.0 on the premise of slightly decreasing output and coal ratio. It indicated that coal B was not suitable for PCI.
(3) When proportion of coal A was used up to 35%, blast furnace smooth operation index (ID_s) raised by 0.08, and carbon content index (ID_t) reduced by over 0.22 with blast furnace output and coal ratio raising. It shows that the coal A was suitable for blast furnace injection.

Acknowledgements Supported by National Key R&D Program of China (2017YFB0304300 & 2017YFB0304302).

References

1. Hua Z, Jianliang Z, Ling B et al (2008) Study on combustion and reactivity of coal powder in blast furnace. Ironmaking 27(1):53–56
2. Minyan W, Daqiang C, Yingru W (2008) Study on the effects of a new additive on coal combustion by TG-DTA. Energy Metall Ind 27(2):32–34
3. Chenglin Q, Jianliang Z, Xinjie H et al (2011) Characteristics of Qingxu coal applied in the 4350 m^3 blast furnace of Taigang. J Univ Sci Technol Beijing 33(01):80–86
4. Shiqiang M, Yubao G, Xuegong B (2009) Mixed spraying of Huangling weak bond coal industry trial. Ironmaking 28(3):56–58
5. Wanren X, Rongren L, Hui Q (2000) Study on utilization of injecting pulverized coal at high IPC ratio on BF at Baosteel. Iron Steel 35(5):5–9
6. Keng W, Boxun Y, Jinming Z et al (2006) Industrial trial of PCI with lean coal on BF at Baosteel. Iron Steel 41(7):21–24

Structural Characterization of the “FeO”-SiO_2 Slags Using Raman Spectra

Yongqi Sun, Yuqi Yao, Mao Chen, and Baojun Zhao

Abstract “FeO”-SiO_2 slags account for a fundamental system in many metallurgical processes and geology, while the slag structures are rarely investigated. In this study, the structures of the quenched “FeO”-SiO_2 slags were quantitatively clarified using the Raman spectra. The samples were equilibrated with metallic iron at high temperatures and quenched rapidly to maintain their structures for examination at room temperature. It was found that as the “FeO” content was 70 wt%, four separate peaks assigned to different SiO_4 tetrahedral appeared at −820, −850, −890, and −950 cm^{-1} in the Raman spectra, which provided a direct clue for the further Raman devolutions. As the “FeO” content increased to 85 wt%, only Q^0(Si) isolations were present in the networks. Raman devolutions showed that with increasing “FeO” content, the mole fraction of Q^0(Si) increased, while those of Q^2(Si) and Q^3(Si) decreased. Furthermore, the relationship between structures and viscosities was discussed, which deepened the knowledge of this fundamental system.

Keywords “FeO”-SiO_2 · Slag structure · Raman · Viscosity

Introduction

Iron silicate or FeO-Fe_2O_3-SiO_2 system is a base for geology [1, 2] and extensive metallurgical processes such as copper smelting [3, 4], copper converting [5], lead processing [6, 7], and slag cleaning [8]. The system is also important in ironmaking and steelmaking processes [9–11]. In these metallurgical processes, their overall efficiencies are greatly determined by the physical properties of high temperature slags such as viscosity, thermal conductivity, heat capacity, liquidus temperature , crystallization behaviours. Currently, there are numerous researches on these physical properties of iron silicate slags for the optimization of various industrial processes [3, 4, 6–9, 11].

Y. Sun · Y. Yao · M. Chen · B. Zhao (✉)
The University of Queensland, Brisbane, Australia
e-mail: baojun@uq.edu.au

B. Li et al. (eds.), *Materials Engineering—From Ideas to Practice: An EPD Symposium in Honor of Jiann-Yang Hwang*, The Minerals, Metals & Materials Series,
https://doi.org/10.1007/978-3-030-65241-8_17

From the fundamental respect, the physical properties of iron silicate slags are determined by their structural orders especially in the short- (2–5 Å) and medium-range (5–20 Å) because generally, the liquid slags and glasses are long-range (>20 Å) disordered [12, 13]. However, there is quite limited research on the structures of this basic system until now, which restricts the further understanding of the natural state as well as the characterization of deep structure–property relationships. Therefore, the present study was motivated to clarify the structures of this system especially the influence of chemical compositions on the short- and medium-range orders.

An iron saturation condition was used in high temperature experiments to control the valence of iron in $FeO\text{-}Fe_2O_3\text{-}SiO_2$ slags , which is generally used for the measurement of physical properties of this system especially the viscosities [14, 15]. The structures of glassy slags after quenching the high temperature liquid slags were identified using Raman spectra at room temperature. Furthermore, viscosities from both experimental measurements and model calculations were selected as a typical physical property here, and the relationship between viscosity and structure was finally discussed briefly.

Experimental

Sample Preparation

As shown in Table 1, three samples with varying chemical compositions, namely "FeO" content, were designed and synthesized using high purity materials, including SiO_2 (99.9 wt%, Alfa Aesar, USA), and "FeO" by the reaction of Fe (99.85 wt%, Goodfellow Cambridge Ltd, UK) and Fe_2O_3 (99.5 wt%, Alfa Aesar, USA). A molybdenum (Mo) crucible was used to hold the sample. To maintain the iron saturation state, around 2 g of iron plate, cut from an iron rod (99.85 wt%, Goodfellow Cambridge Ltd, UK), was placed at the bottom of the Mo crucible. A vertical tube furnace was used to melt the slags at 1500 °C under the atmosphere of ultrahigh purity Ar gas (99.999 vol%), which can realize the temperature accuracy within 3 °C, as detailed in the previous studies [16–19]. After melting and equilibration, the high temperature liquid slags were rapidly dropped into cold water to obtain a glassy state. Afterwards, the glassy samples were dried, crushed, mounted, and polished for the further analysis.

Table 1 Chemical composition of liquid slags in saturation with metallic iron (wt%)

Sample	"FeO"	"SiO_2"	Fe^{2+}/TFe* (%)
FS-1	60	40	99.56
FS-2	70	30	98.69
FS-3	85	15	94.77

*calculated by FactSage 7.3 [20] at 1500 °C

Structural Characterization of the Prepared Glasses

The chemical compositions of the quenched samples were measured by Electron Probe X-ray Micro-analysis (EPMA), and it was proved that the content of MoO_3 was less than 0.5 wt%. The structures of the obtained glassy samples were then characterized using Raman spectra where a Renishaw Raman spectrometer was applied with the excitation wavelength of 514 nm within the Raman shift of 200–2000 cm^{-1} at room temperature, and the light source of a 1 mw semiconductor.

Based on the calculations using FactSage 7.3 [20], the mole ratio of Fe^{2+} in the total iron (TFe) for all samples was found to be higher than 94.5 wt% at 1500 °C under the iron saturation condition, as detailed in Table 1. The $FeO\text{-}Fe_2O_3\text{-}SiO_2$ slags can be represented as "FeO"-SiO_2 system. In addition to the experimental measurements, FactSage 7.3 was also used to calculate the viscosities of the high temperature slags for the further discussion of structure–viscosity relationship.

Results and Discussion

Raman Spectra of Different Glasses

To clarify the structure of slags, Raman spectra were first collected, as shown in Fig. 1. As can be observed, the total Raman spectra were divided into three main Raman shift regions: the low Raman shift region (400–600 cm^{-1}) assigned to T-O-T bending vibrations (T represents Si and Fe), the intermediate Raman shift region (600–800 cm^{-1}) assigned to Si–O symmetry stretching vibrations, and the high Raman shift region (800–1150 cm^{-1}) assigned to stretching vibrations of SiO_4 tetrahedral [21–24].

Overall, with increasing "FeO" content, the band located at approximately 700 cm^{-1} assigned to Si–O symmetry stretching vibrations became relatively more pronounced; while the relative intensity of SiO_4 vibrations (800–1150 cm^{-1}) became less prominent due to the decrease of SiO_2 content in the slags. Moreover, the centre of SiO_4 vibrations band (800–1150 cm^{-1}) considerably shifted to the lower Raman shift range with the increase of "FeO" content, which qualitatively indicated that the degree of polymerization (DOP) of the networks decreased.

In particular, it can be observed that, there were four separate peaks in the Raman spectra of sample FS-2, positioned at −820, −850, −890, and −950 cm^{-1}, respectively. These peaks originated from the stretching vibration of four various SiO_4 tetrahedral, denoted as $Q^i(Si)$, i = 0, 1, 2 and 3, where i represents the number of bridging oxygen (BO) per SiO_4 tetrahedral. This also provided an excellent and direct clue to further deconvolute the Raman spectra for quantitative analysis. In addition, the relative intensity of −820 cm^{-1} peak, attributed to $Q^0(Si)$, was remarkably enhanced with increasing "FeO" content, while the band near 950 cm^{-1}, attributed to $Q^3(Si)$, considerably decayed, which also demonstrated a lower DOP of the glass networks with increasing "FeO" content.

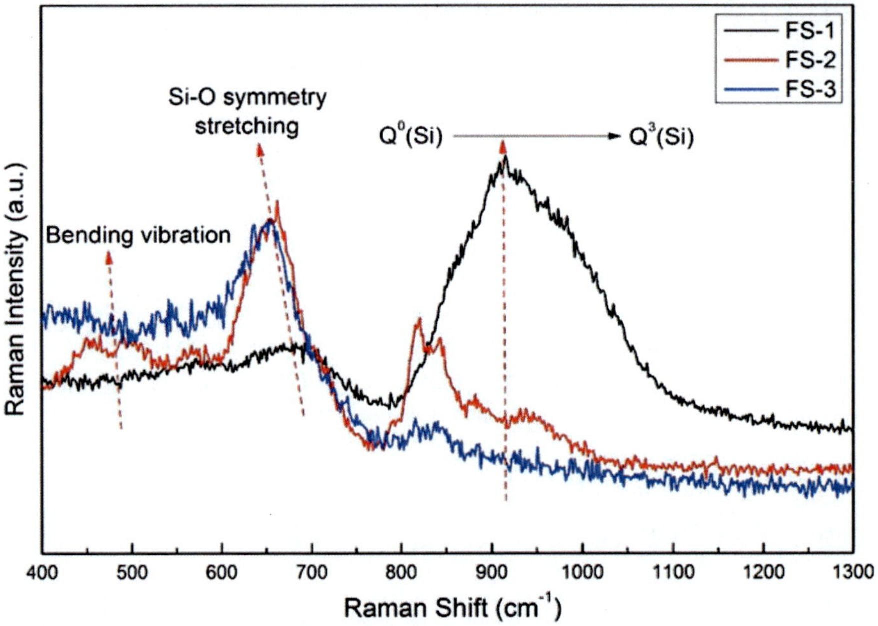

Fig. 1 Raman spectra of the prepared glassy samples. (Color figure online)

Deconvolution of Raman Spectra of Different Glasses

To quantitatively identify the silicate structure especially the concentrations of various SiO_4 tetrahedral, deconvolution of the Raman spectra using four Gaussian functions was carried out in the Raman shift range of 800–1150 cm^{-1} based on three fundamental principles, i.e., the observed peak and bond distributions in the Raman spectra, the previous researches on silicate glasses [21–24] and the mathematical optimizations. The fitting process is presented in Figs. 2, 3, and 4, representing samples FS-1, FS-2, and FS-3, respectively. As can be noted from Figs. 2 and 3, four Gaussian peaks can well fit the Raman spectra of samples FS-1 and FS-2 with high SiO_2 content, in the Raman shift range of 800–1150 cm^{-1}, which also demonstrated the reasonability of the present methodology based on peak/bond distributions and Gaussian fitting.

However, for sample FS-3 with the "FeO" content further increasing to 85 wt%, Fig. 4 shows that the silicate networks got markedly depolymerized, and therefore, only Q^0(Si) isolations were excited at −820 cm^{-1}. This was in consistence with that, with increasing "FeO" content, the relative intensity of Si–O symmetry stretching vibrations (600–800 cm^{-1}) got enhanced because with higher "FeO" content of 85 wt%, Q^0(Si) isolations became the dominant structural unit, which showed a typical Si–O symmetric local structure.

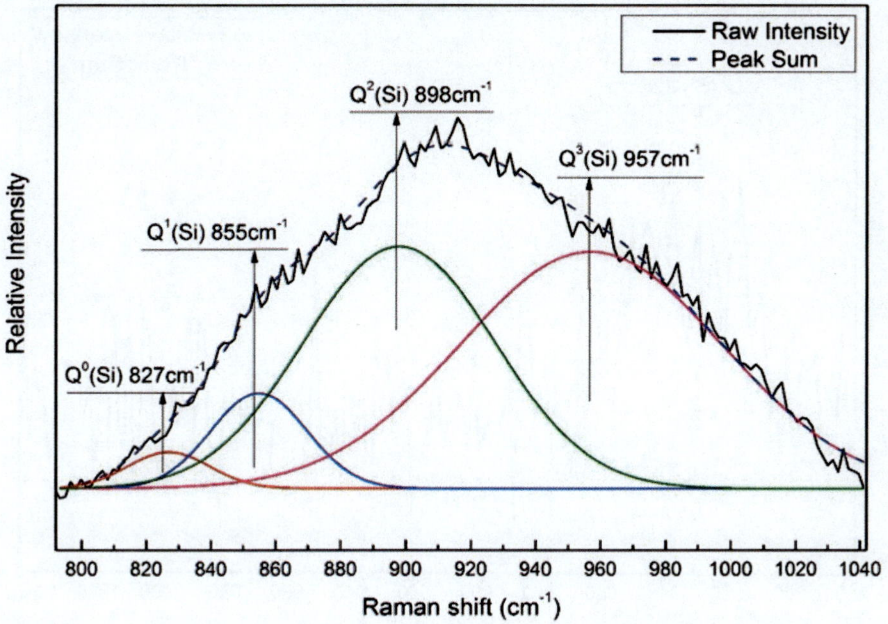

Fig. 2 Deconvolution of the Raman spectra of sample FS-1. (Color figure online)

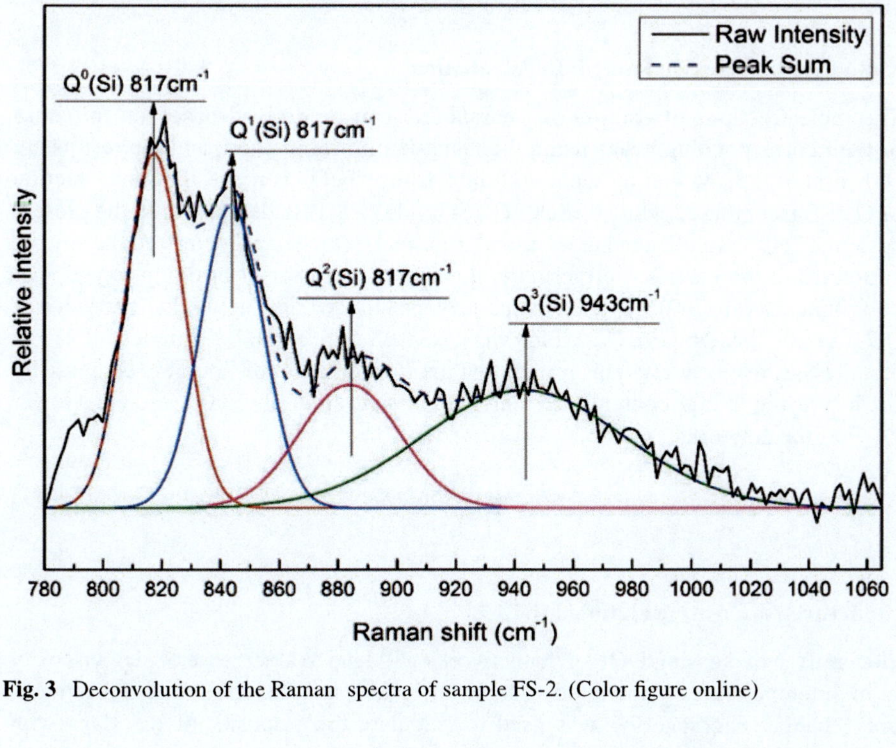

Fig. 3 Deconvolution of the Raman spectra of sample FS-2. (Color figure online)

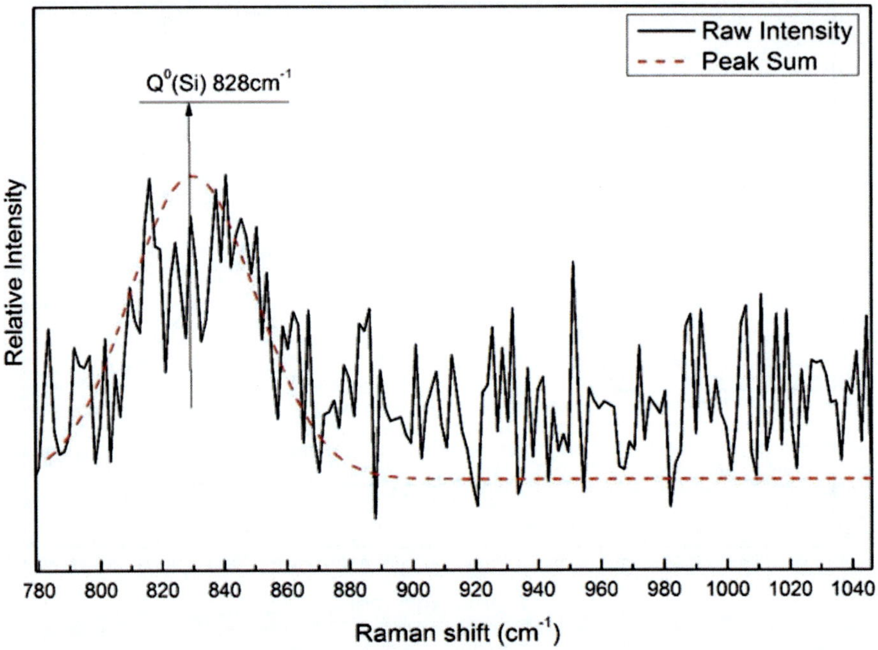

Fig. 4 Deconvolution of the Raman spectra of sample FS-3. (Color figure online)

Mole Fractions of Various SiO_4 Tetrahedral

The mole fractions of each $Q^i(Si)$ tetrahedral can be derived based on the areas of their corresponding peaks using the Gaussian fitting method, and the results are shown in Fig. 5. As can be seen, with increasing "FeO" content, the mole fraction of $Q^0(Si)$ is increased while those of $Q^2(Si)$ and $Q^3(Si)$ are decreased. In the present system, "FeO" overall acted as a network modifier, which depolymerized the silicate structures in the glasses. Furthermore, the average number of bridging oxygen per coordinated atom ($\overline{BO}$) was used to quantify the DOP of slag networks. The value of $\overline{BO}$ was calculated by Eq. (1), which were 2.389, 1.525, and 0 for samples FS-1, FS-2, and FS-3, respectively. This proves that the $\overline{BO}$ contents considerably decrease by the increasing "FeO" content, i.e., a higher concentration of "FeO" leads to a lower DOP of the networks.

$$\overline{BO} = \sum_i iQ^i(Si) \tag{1}$$

Structure–Viscosity Relationship

Generally, a decreasing DOP of the networks will lead to a decrease of slag viscosity at high temperatures if the bond strength of local structural units does not remarkably change. FactSage 7.3 was used to calculate the viscosity of the slags with

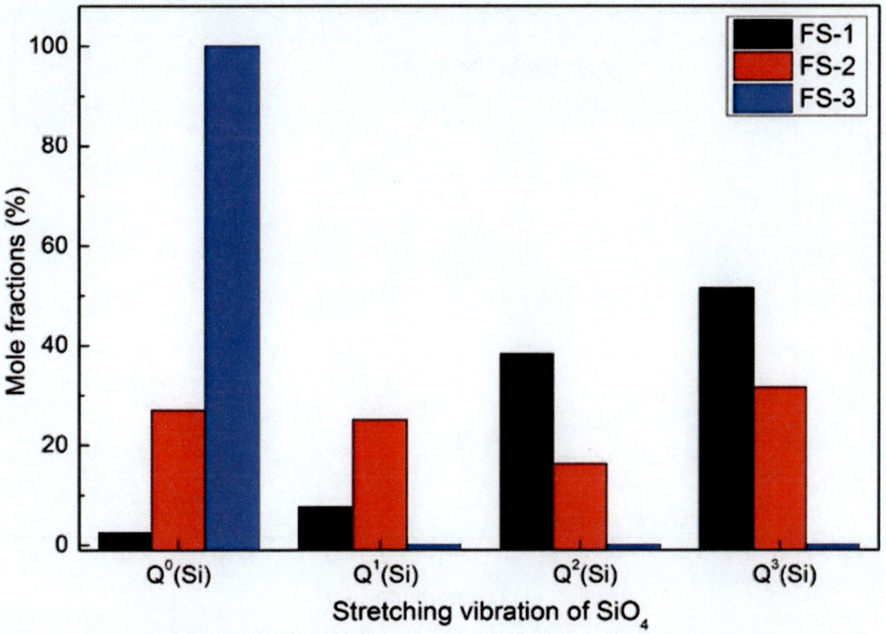

Fig. 5 Mole fractions of various Q^i(Si) in the "FeO"-SiO_2 glass system. (Color figure online)

different "FeO" contents as shown in Fig. 6. As expected, the viscosity decreases with increasing "FeO" concentration which is in agreement with the experimental results in our previous study [25]. In detail, sample FS-1 with 60 wt% "FeO" has the highest viscosity, and sample FS-3 with 85 wt% "FeO" had the lowest viscosity with a good consistence between the experimental and calculated results. It can be directly deduced here that "FeO" was a network modifier, i.e., a higher "FeO" content depolymerized the slags' silicate network more prominently, and thus decreased the viscosity of slags at high temperatures.

Conclusions

In summary, "FeO"-SiO_2 accounts for a significant and fundamental system in extensive industrial processes and geology. In this study, the structures of this system were identified using Raman spectra. It was found that, as the "FeO" content increased, the mole fraction of Q^0(Si) increased while those of Q^2(Si) and Q^3(Si) decreased. In particular, as the "FeO" content further increased to 85 wt%, only Q^0(Si) isolations are present. Based on the deconvolution of Raman spectra, the value of bridging oxygen per coordinated silicon atom was calculated, and finally the relationship between structures and viscosities is discussed briefly.

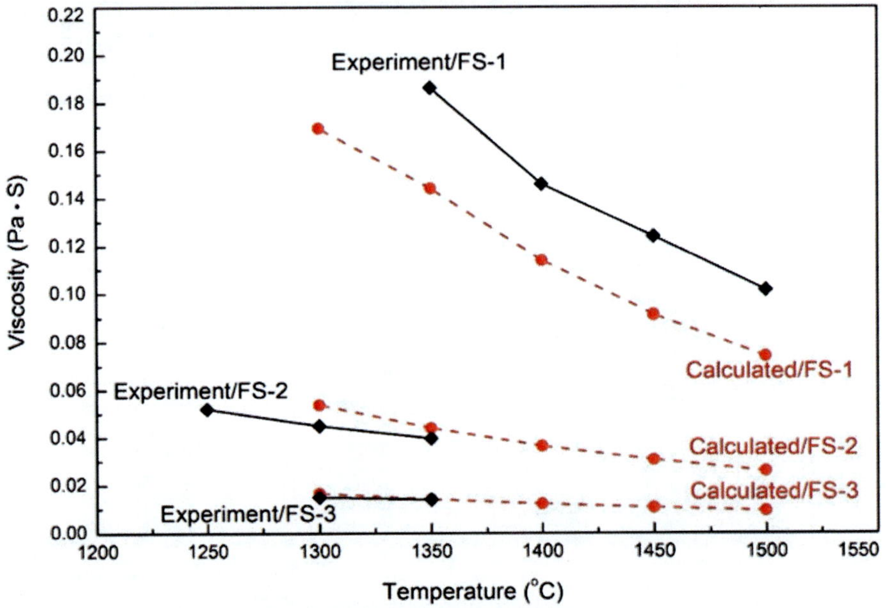

Fig. 6 Viscosity versus temperature of the "FeO"-SiO_2 system. (Color figure online)

Acknowledgements The authors acknowledge the facilities, and the scientific and technical assistance, of the Australian Microscopy & Microanalysis Research Facility at the Centre for Microscopy and Microanalysis, The University of Queensland.

References

1. Rasmussen B, Meier DB, Krapež B, Muhling JR (2013) Iron silicate microgranules as precursor sediments to 2.5-billion-year-old banded iron formations. Geology 41:435–438
2. Buss HL, Lüttge A, Brantley SL (2007) Etch pit formation on iron silicate surfaces during siderophore-promoted dissolution. Chem Geol 240:326–342
3. Busolic D, Parada F, Parra R, Sanchez M, Palacios J, Hino M (2011) Wastes as resources: update on recovery of valuable metals from copper slags. Miner Process Ext M 120:32–36
4. Park HS, Park SS, Sohn I (2011) The viscous behavior of FeO_t-Al_2O_3-SiO_2 copper smelting slags. Metall Mater Trans B 42:692–699
5. Altundoğan HS, Tümen F (1997) Metal recovery from copper converter slag by roasting with ferric sulphate. Hydrometallurgy 44:261–267
6. Kudo M, Jak E, Hayes P, Yamaguchi K, Takeda Y (2000) Lead solubility in FeO_x-CaO-SiO_2 slags at iron saturation. Metall Mater Trans B 31:15–24
7. Barna R, Moszkowicz P, Gervais C (2004) Leaching assessment of road materials containing primary lead and zinc slags. Waste Manage 24:945–955
8. Heo JH, Kim BS, Park JH (2013) Effect of CaO addition on iron recovery from copper smelting slags by solid carbon. Metall Mater Trans B 44:1352–1363
9. Mills KC, Sridhar S (1999) Viscosities of ironmaking and steelmaking slags. Ironmak Steelmak 26:262–268

10. Nath SK, Kumar S (2013) Influence of iron making slags on strength and structure of fly ash geopolymer. Constr Build Mater 38:924–930
11. Wang W, Lu B, Xiao D (2016) A review of mold flux development for the casting of high-Al steels. Metall Mater Trans B 47:384–389
12. Anderson PW (1995) Through the glass lightly. Science 267:1609–1618
13. Elliott SR (1991) Medium-range structural order in covalent amorphous solids. Nature 354:445–452
14. Shiraishi Y, Ikeda K, Tamura A, Saito T (1978) On the viscosity and density of the molten FeO–SiO_2 system. Trans JIM 19:264–274
15. Ji F, Du S, Seetharaman S (1997) Experimental studies of the viscosities in the CaO-Fe_nO-SiO_2 slags. Metall Mater Trans B 28:827–834
16. Sun Y, Chen M, Zhao B (2019) Modification of the structures of CaO-SiO_2-Al_2O_3-MgO slags using various minor elements. J Non-Cryst Solids 515:50–57
17. Chen M, Sun Y, Balladares E, Pizarro C, Zhao B (2019) Experimental studies of liquid/spinel/matte/gas equilibria in the Si-Fe-O-Cu-S system at controlled P (SO_2) 0.3 and 0.6 atm. Calphad 66: 101642
18. Sun Y, Chen M, Cui Z, Contreras L, Zhao B (2020) Phase equilibrium studies of iron silicate slag under direct to blister copper-making condition. Metall Mater Trans B 51:1–5
19. Sun Y, Chen M, Cui Z, Contreras L, Zhao B (2020) Phase equilibrium studies of iron silicate slag in the liquid/spinel/white metal/gas system for copper converting process. Metall Mater Trans B 51:426–432
20. Bale CW, Bélisle E, Chartrand P, Decterov SA, Eriksson G, Gheribi AE, Hack K, Jung IH, Kang YB, Melançon J, Pelton AD (2016) FactSage thermochemical software and databases, 2010–2016. Calphad 55:1–19
21. Neuville DR, de Ligny D, Henderson GS (2014) Advances in Raman spectroscopy applied to earth and material sciences. Rev Mineral Geochem 78:509–541
22. Hwa LG, Hwang SL, Liu LC (1998) Infrared and Raman spectra of calcium alumino–silicate glasses. J. Non-Cryst Solids 238:193–197
23. Daniel I, Gillet P, Poe BT, McMillan PF (1995) In situ high-temperature Raman spectroscopic studies of aluminosilicate liquids. Phys Chem Miner 22:74–86
24. Mysen BO (1996) Phosphorus speciation changes across the glass transition in highly polymerized alkali silicate glasses and melts. Am Mineral 81:1531–1534
25. Chen M, Raghunath S, Zhao B (2013) Viscosity measurements of "FeO"-SiO_2 slag in equilibrium with metallic Fe. Metall Mater Trans B 44:506–515

Removal of Rare-Scattered Metal Impurities in Zinc Sulfate Solution by Ozone Oxidation

Zhang Lihua, Luo Yaoyao, Wang Tian, Qu Hongtao, Zhang Jiale, Zhang Libo, and Luo Yongguang

Abstract There are abundant strategic rare-scattered metal resources in lead and zinc deposits in China, especially zinc resources in Yunnan province, which are associated with indium (In), selenium (Se), tellurium (Te), etc. Small amounts of these rare-scattered metal ions are dissolved into zinc acidic leaching solutions, and are difficult to remove by conventional zinc powder replacement. They are enriched in the $ZnSO_4$ solution, which causes serious problems to the zinc electrowinning process such as hydrogen generation, zinc re-dissolution, plate burning, low current efficiency, etc. So, they need to be deeply purified from $ZnSO_4$ solution. This article used ozone, a strong oxidizing agent, to make oxidative precipitation of the impurities from $ZnSO_4$ solution. The effects of reaction temperature, ozone flow rate, and residence time on In, Se, and Te removal together with the associated Mn removal and Zn loss were investigated. The ozonation reaction kinetics was evaluated by plotting the negative logarithm of the metals in solution concentration as a function of reaction time. One precipitate prepared at optimized conditions was characterized by XRD, EDS, and SEM to identify the chemical compounds and morphology.

Keywords Zinc hydrometallurgy · Ozone oxidation · Rare scattered metal impurities · Deep purification · Zinc sulfate solution

---

Z. Lihua · L. Yaoyao · W. Tian · Z. Jiale · Z. Libo (✉) · L. Yongguang (✉)
State Key Laboratory of Complex Nonferrous Metal Resources Clean Utilization, Kunming University of Science and Technology, Kunming 650093, Yunnan, China
e-mail: zhanglibopaper@126.com

L. Yongguang
e-mail: ynchlyg@163.com

Key Laboratory of Unconventional Metallurgy, Ministry of Education, Kunming 650093, Yunnan, China

Faculty of Metallurgical and Energy Engineering, Kunming University of Science and Technology, No. 68 Wenchang Road, 121 Street, Kunming 650093, Yunnan, China

Q. Hongtao · L. Yongguang
Yunnan Chihong Zn & Ge Co., LTD., Qujing 655011, Yunnan, China

B. Li et al. (eds.), *Materials Engineering—From Ideas to Practice: An EPD Symposium in Honor of Jiann-Yang Hwang*, The Minerals, Metals & Materials Series,
https://doi.org/10.1007/978-3-030-65241-8_18

Introduction

According to geological surveys, there are abundant strategic rare-scattered metal (RSM) resources in lead and zinc deposits in China, especially zinc resources in Yunnan province, which are associated with indium (In), selenium (Se), tellurium (Te), etc. [1, 2]. In the sulfuric acid-leaching process of zinc concentrates, these RSMs dissolve into the solution with zinc and other desired metals. Most of them in high valent state are subsequently hydrolyzed and precipitated into the residues when the acidity of the post-leaching solution reduced, but still small amounts in the low valent form remain in the leachate solution and are accumulated when the zinc electrolytes are recirculated. A certain concentration of the RSM impurities in zinc sulfate solution has a deleterious effect on the downstream electrowinning process. Similar to other conventional electropositive impurities, such as Cu, Cd, Ni, Co, etc., in the electrolyte, these RSM impurities also affect the cathodic current efficiency (CE), electrical power consumption (PC), cell voltage (CV), and deposit quality of Zn product [3, 4]. Therefore, it is necessary to search an appropriate method to remove them in a purification step prior to zinc electrowinning process.

As RSM ions of In, Se, and Te are more electropositive than zinc, there should not be any problem to remove them by adding zinc powder during zinc sulfate solution purification in theoretical. However, in practice it is still a challenge due to the limited reaction kinetics caused by the formation of passivating layers on the surface of Zn powder, and the potential re-dissolution of the precipitates when the zinc particles are completely consumed. For example, in a zinc hydrometallurgy plant in Yunan, the amounts of Se and Te in the zinc sulfate solution increased from 1.5 and 3.8 mg/L in 2010 to 3.5 and 6.7 mg/L in 2013, and correspondingly the average CE of the long period electrowinning sharply decreased from 86.7% to 82.6%, respectively, and sometimes even below 78% [3].

Some methods including multi-elements alloy hydrolytic adsorption [3, 5] and activated carbon adsorption [6] are studied to remove In, Se, or Te from solutions. However, for the former the removals of Se and Te are only 39% and 23% at optimized parameters, respectively [3]. In addition, new alloy impurity ions should be introduced into the system and cause negative effect for the downstream electrowinning process. For the latter method, the removing efficiencies of the RSMs are also not very efficient, and it needs complicated regeneration technology of activated carbon. Another established process for the purification of zinc sulfate solutions is the use of different oxidants for oxidative precipitation of the impurity metal ions. The most commonly used oxidizing agents in industry are MnO_2 and $KMnO_4$ which afford high standard redox potential of $E^{\ominus} = 1.23$ and 1.67 V, respectively. At present, the RSM ions in low valent can be partially oxidized into higher valent ions by MnO_2 and/or $KMnO_4$ and then hydrolytically precipitated from zinc sulfate solutions into the residues. However, they are creating a manganese balance problem in the electrowinning process. Therefore, it is essential to explore an alternative oxidizing agent for the deep purification of the metallic impurities to the desired low levels and does not introduce any new impurities into the system.

Ozone (O_3) is well known as a strong oxidant with a high redox potential of 2.07 V, which enables to oxidize all metals and their compounds in solution except platinum, gold, and iridium [7, 8]. Its strong oxidizing effectiveness is based on the multiple effects produced by the oxidative activity of ozone and ozone-derived oxidizing species such as free hydroxyl radicals ·OH [9, 10]. On the other hand, while ozone itself rapidly decomposes in water, leaves oxygen as the only residual leading to no secondary pollution. Therefore, there has been considerable interest in using ozone application in environment and hydrometallurgical processes. Especially in hydrometallurgy, it has been widely reported in the literatures that O_3 is used as a promising alternative oxidant to make metal ions at low valence oxidize into that at high valent state, and then hydrolyzed as precipitations.

Calzado et al. [11] reported the use of ozone to recover Ni(II) from sulfate solutions. By strictly controlling a minimum pH of about 6.8 at room temperature, the first product of $Ni(OH)_2$ was precipitated, and then was oxidized by ozone to produce β-NiOOH which had a higher density and settleability than $Ni(OH)_2$. The oxidation reaction at pH 6.5 followed a first order rate: $2Ni^{2+} + O_3 + 3H_2O \rightarrow 2NiOOH + 4H^+ + O_2$. On the other hand, Nishimura and Umetsu [12] found that by controlling low pH of cobalt and nickel sulfate solution in the range of 2.5–5.0 at 60 °C, ozone oxidation was very effective to separate cobalt ions selectively from nickel sulfate solutions. The reason was that the oxidation of nickel ion was extremely slow in comparison with cobalt. Tian et al. [13] studied the oxidation-precipitation kinetics of cobalt(II) from cobalt chloride solution by ozone. The authors indicated that the precipitation reaction was diffusion-controlled, and the oxidation rate increased obviously with the increase of stirring speed but was independent of the initial concentration or solution temperature. They also found the linear relationship between ozone partial pressure or flow rate and oxidation rate. The oxidation reaction using ozone as the precipitant was: $2Co^{2+} + O_3 + 3H_2O \rightarrow 2CoOOH + 4H^+ + O_2$. Nishimura and Umetsu [14] also investigated the removal of arsenic with manganese from dilute acidic sulfate solutions by oxidation-precipitation using ozone. They found that by formation of manganese (III) arsenate and uptook of arsenic by manganese dioxide formed by ozone oxidation, arsenic entities were removed to a sufficiently low concentration (below the limit of 0.1 mg/L) in the pH range 1.0–3.0 at 25 °C, where co-precipitation with ferric hydroxide is not very effective to remove arsenic. Lin [15] studied the ozonation kinetics of iron and manganese in aqueous solution and found that a zero-order kinetic model can be used to describe the ozonation process. The chemical reaction involved in the oxidation of iron and manganese ions with ozone can be presented as: $4Fe^{2+}{}_{(aq)} + 4O^3 + 10H_2O \rightarrow 4Fe(OH)_{3(s)} + 5O_2 + 8H^+$, and $Mn^{2+}{}_{(aq)} + O_3 + 2OH^- \rightarrow MnO_{2(s)} + O_2 + H_2O$. Based on the above studies, we can understand that ozonation is an effective means to oxidize metal ions with low valence into high valence and then for hydrolytic precipitation at a very low limit amounts of the metals in the solutions.

In the current work, oxidative precipitation using ozone was explored for removal of In, Se, and Te together with the associated Mn and Zn from zinc sulfate solution and for identification the forms of precipitates morphology obtained under predetermined conditions. Optimization of the most effective parameters on the impurities removal was also studied.

Materials and Methods

In the experiments, the $ZnSO_4$ solution derived from a zinc extraction plant in Yunnan province of China which was the vitriol supernatant after Zn powder replacement process. Elemental analysis result of the solution is presented in Table 1.

The experimental setup used for ozonation experiments was schematically shown in Fig. 1. Ozone gas was generated using a commercial ozone generator (CFZY-6B, Hangzhou Rongxin Electronic Equipment Co., Ltd., China) which the oxygen enriched dry air was used as feed gas. Inside the ozone generator, the oxygen was ionized to produce ozone under an electric field with high voltage and frequency in a discharge tube. The flow rate of ozone produced by this apparatus was determined using iodometric method, which varied within the range from 1.7 to 8.6 g/h of O_3 and was dependent on the flow of oxygen enriched air from 25 to 125 L/h. Ozonation experiments were performed in a sealed glass cylinder with a 4-port polyethylene cap, and 500 mL $ZnSO_4$ solution was charged for each experiment. The ozone gas was sparged into the $ZnSO_4$ solution using a fish tank aerator with 5 cm in diameter to ensure generation of small bubbles. The gas and liquid phases were mixed at 500 rpm on an electric stirrer. A thermocouple was immersed in the solution to actually control the 20–90 °C reaction temperature with ±0.2 °C sensitivity. The residual ozone from

Table 1 Main chemical compositions of $ZnSO_4$ solution before ozonation process

Element	Zn	Mn	Se	Te	In	pH
Content	146 g/L	4.17 g/L	1.022 mg/L	0.36 mg/L	3.87 mg/L	5.2

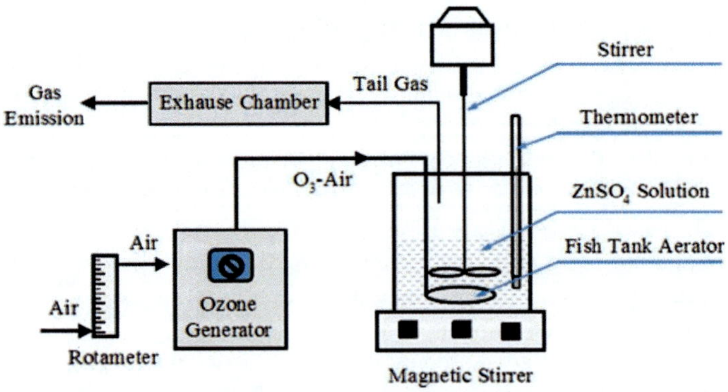

Fig. 1 Schematic diagram of the ozonation experimental setup. (Color figure online)

the outlet was absorbed by the ozone destructor equipped with the ozone generator by a potassium iodide solution before discharge to the atmosphere. At the end of a test, ozone gas was closed and slurry was then separated using a Buchner funnel and vacuum pump at 20 psig pressure. Experiments were carried out to study the effects of reaction temperature, ozone flow rate, and reaction time on the removals of In, Se, and Te and together with the removal of Mn and loss of Zn.

In, Se, Te, and Mn in the solutions before and after each experiments were determined by using an induced coupled plasma atomic emission spectrometry (ICP-MS, 7700x, Agilent Technologies, Inc., USA), and Zn concentration in the solutions was analyzed by complexometric titration using disodium ethylene diamine tetra acetate (Na-EDTA) as the titrant and xylenol orangs as the indicator. Distilled water was used to make up all required solutions. X-ray diffraction (XRD, Rigaku D/Max 2200) was performed to analyze the phase composition of the precipitations. The morphology and element analysis of the precipitations were examined by scanning electron microscope (SEM, XL-30E) and energy dispersive X-ray spectroscopy (EDX), respectively.

The percentage removal of RSMs impurities and Mn or loss of Zn from zinc sulfate solution was calculated by the following formula:

$$\%\ \text{Removal/Loss} = \frac{C_{M,O} - C_{M,t}}{C_{M,O}} \times 100 \tag{1}$$

where $C_{M,o}$ and $C_{M,t}$ refer to the metal ion concentrations before and after ozonation treatment (g/L), respectively.

Results and Discussion

Effect of Temperature

A series of experiments was carried out to evaluate the effect of temperature on the removals of In, Se, and Te by varying the reaction temperature in the range of 20–90 °C with ozone flow rate of 3.1 g/h (corresponding to the O_2 flow rate of 50 L/h) and reaction time of 10 min. The obtained results are indicated in Fig. 2a. On the other hand, in zinc hydrometallurgy, the $ZnSO_4$ solution usually contains approximately 3–5 g/L Mn^{2+}, which is introduced during removal of ferrous iron using pyrolusite and parts is from the acid leach liquor of zinc concentrates with manganese [10]. The Mn^{2+} ions in $ZnSO_4$ solutions are also easily oxidized by ozone to form precipitates of MnO_2 and/or $ZnMn_2O_4$, and/or ions Mn^{3+} or MnO_4^{-1}, etc. [10, 14, 15]. Therefore, it's necessary to explore the removal of Mn and the loss of Zn on O_3 ozonation process, and the result is shown in Fig. 2b.

From Fig. 2a, b, it is shown that reaction temperature is clearly important in the ozonation process, as the removals of In, Se, and Te increase with increasing

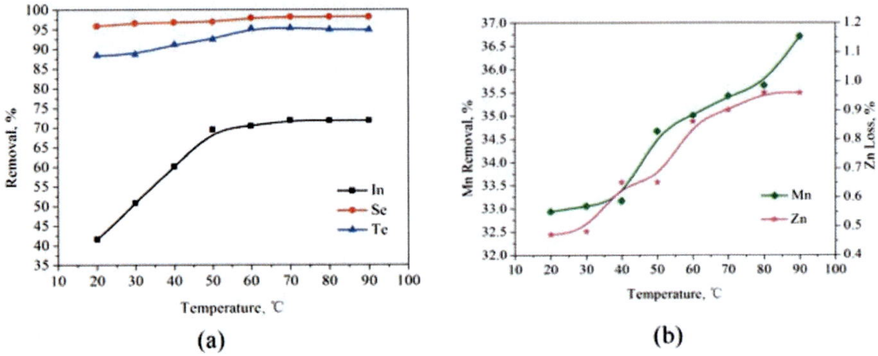

Fig. 2 Effects of reaction temperature on **a** the percentage removals of In, Se, and Te, and **b** the percentage removal of Mn and the loss of Zn. (Color figure online)

temperature to 60 °C and then reach a plateau at higher temperature, whereas the removal of Mn and the loss of Zn slightly increase with increasing temperature. This is found to be related to the temperature dependence of not only the reaction rate but also the ozone solubility in solution. As we know, the reaction rate grows rapidly with operating temperature, which is beneficial to agglomeration and growth of precipitate particles. However, on the other hand, the amount of dissolved ozone decreased with increasing temperature [16, 17]. It is therefore important to provide an optimal balance of gas-liquid mass transfer rate and ozone solubility to achieve maximum removal rates of In, Se, Te, and Mn, but minimum loss of Zn. So a temperature of 60 °C seems to be favorable, whereas the maximum removals of In, Se, Te, and Mn are obtained about 70.4%, 97.9%, 95.2%, and 35.0%, and the loss of Zn are 0.86%, respectively.

Effect of Ozone Flow Rate

The effect of ozone flow rate on RSMs impurities and Mn removals and Zn loss was investigated at reaction time of 10 min and temperature of 60 °C. The results are presented in Fig. 3a, b.

As shown in Fig. 3a, it is indicated that increasing the ozone mass flow rate from 1.7 to 4.6 g/h resulted in substantial increase in the removals of In, Se, and Te, but higher ozone does not seem to have a marked effect. The removal of Mn and the loss of Zn increase slightly with increasing of ozone flow rate. Considering the efficiency and cost of the ozone generation, 4.6 g/h ozone is considered to be the optimal amount, and applied in the subsequent experiments.

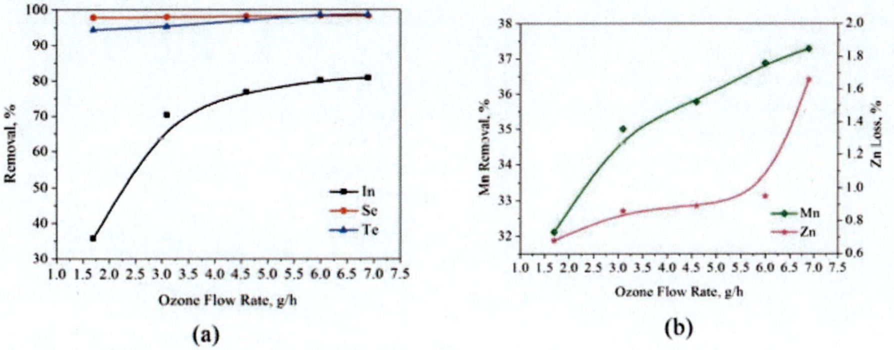

Fig. 3 Effects of ozone flow rate on **a** the removals of In, Se, and Te, and **b** the removal of Mn and the loss of Zn. (Color figure online)

Effect of Reaction Time

The effect of reaction time on removals of the metals was evaluated by varying the operating time in the range of 1–30 min with ozone concentration of 4.6 g/h and reaction temperature of 60 °C. The obtained results are illustrated in Fig. 4a, b.

As seen in Fig. 4a, b, it is indicated that most of the In, Se, Te and Mn remove expeditiously in the first 10 min, then became slow gradually and remained almost unchanged after 20 min. The Zn loss monotonically increases from 0.4 to 1.7% with increasing reaction time from 1 to 30 min. The lasting reaction time over 10 min does not seem to have a noticeable effect for the removals of In, Se, and Te, but aggravate the loss of Zn and the consumption of ozone. Therefore, the optimum time appeared to be 10 min. At this time, 80.2% of In, 98.2% of Se, 98.7% of Te, and 36.9% of Mn are removed from the zinc sulfate solution, while the loss of Zn is 0.95%, respectively.

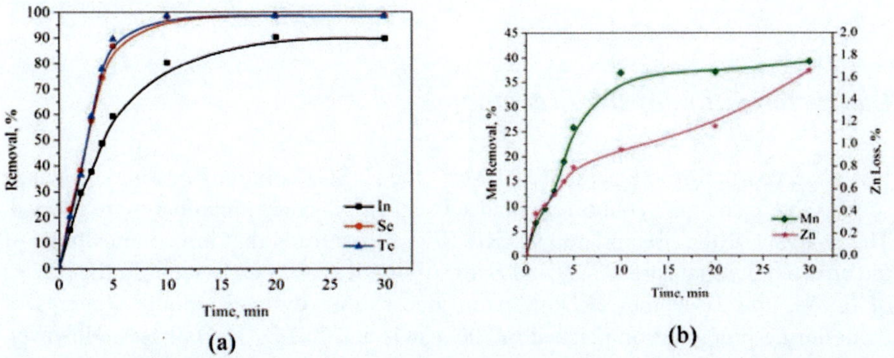

Fig. 4 Effects of time on **a** the removals of In, Se, and Te, and **b** the removal of Mn and the loss of Zn. (Color figure online)

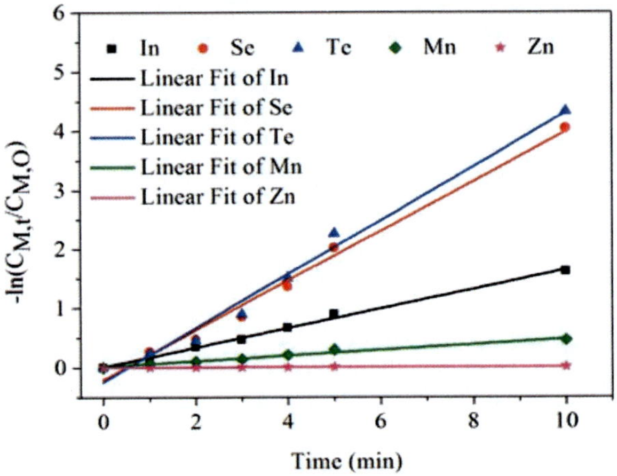

Fig. 5 The negative logarithm of the ratio of metals concentration ($C_{M,t}/C_{M,O}$) as a function of time. (Color figure online)

Table 2 Line equations and R^2 values for Fig. 5

Metal	Equation	R^2
In	y = 0.1633x + 0.0123	0.9954
Se	y = 0.4189x − 0.2071	0.9858
Te	y = 0.4543x − 0.2414	0.9815
Mn	y = 0.0465x + 0.0166	0.9701
Zn	y = 0.00084x + 0.0024	0.7769

The kinetics of the ozonation reactions are further evaluated by plotting the negative logarithm of the metals in solution concentration as a function of reaction time, as shown in Fig. 5, and the constants and quality of fit indicated in Table 2. It can be seen that a straight line for a first order reaction is obtained for the oxidative removal of the metals in the $ZnSO_4$ solution, which indicated the oxidative reaction is diffusion-controlled as the similar kinetics of oxidation of Co(II) by ozone [13].

Characterization of the Precipitates

It is observed that in the presence of Mn^{2+} the $ZnSO_4$ solutions become brown as soon as the ozone gas is bubbled in and subsequently brown precipitates are formed. The analysis of the precipitate by XRD (Fig. 6) confirms that the compositions of the brown precipitate are MnO_2 and $ZnSO_4{\cdot}H_2O$. It is difficult to identify the phases of In, Se, and Te in the XRD spectrum due to the so small amount contents of them during precipitation masked by the MnO_2 and $ZnSO_4{\cdot}H_2O$ phases. However, characterization of the precipitate with EDS confirms the presence of these RSMs (Fig. 5b).

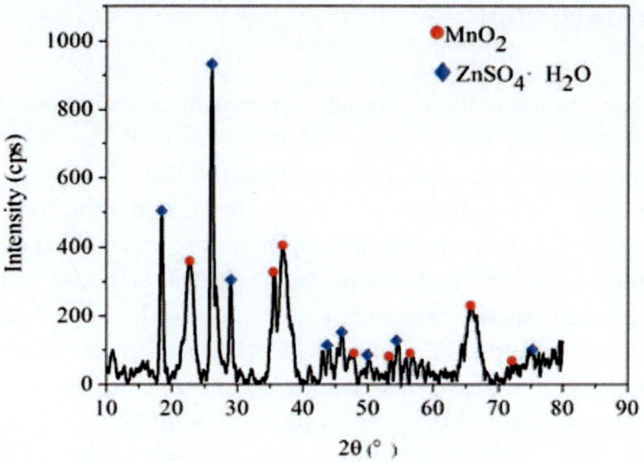

Fig. 6 X-ray diffraction pattern of the precipitates at temperature of 60 °C, ozone concentration of 3.75 g/L, and time of 10 min. (Color figure online)

Therefore, the O_3 oxidative removals of RSM ions is accompanied with a simultaneous oxidation of Mn^{2+}, and a reaction mechanism is proposed as follows:

$$Mn^{2+} + O_3 + H_2O = MnO_2 + O_2 + 2H^+ \quad (2)$$

For the loss of Zn, it forms as $ZnSO_4{\cdot}H_2O$ and is absorbed by MnO_2 and co-precipitated into residues. For the In, Se, and Te, the assay is not sufficient to distinguish their chemical compounds in the residues and more detailed chemistry will be attempted in the future.

Figure 7a presents SEM image of the precipitate. It is determined that the compound precipitates as bulks with different sizes, which is good for sedimentation and filtration operation.

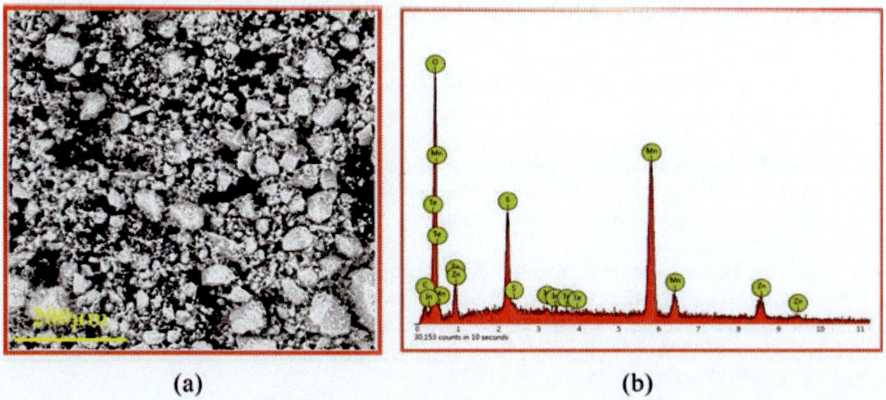

Fig. 7 **a** SEM and **b** EDS images of the precipitates at temperature of 60 °C, ozone concentration of 3.75 g/L, and time of 10 min. (Color figure online)

Conclusions

The removals of In, Se, and Te from zinc sulfate solution by ozone oxidation was investigated with focuses on the effects of various process factors including the temperature, O_3 flow rate, and reaction time. In, Se, and Te removals should be up to 80.2%, 98.2%, and 98.7%, respectively, under the conditions of reaction temperature of 60 °C, O_3 flow rate of 4.6 g/h, and reaction time of 10 min, while the associated removal of Mn and loss of Zn are 36.9% and 0.95%, respectively. The results from reaction kinetics indicate that the ozone oxidation process is a first order reaction and diffusion-controlled. The XRD analysis of the precipitate show that the main components of the precipitate are MnO_2 and $ZnSO_4 \cdot H_2O$, and EDS analysis confirms the presence of RSMs in the residue. SEM image illustrates that the compound precipitated as bulks with different sizes, which is good for sedimentation and filtration. However, chemical compounds of RSMs in the residues can not be sufficient to distinguish at present and more detailed chemical reaction and removal mechanism will be attempted in the future.

Acknowledgements The authors are grateful for Yunnan Chihong Zn & Ge Co., Ltd Nonferrous Metal Electrodeposition Technology Innovation Team of Yunnan Province (201905E160007) and the financial support of the Talent training project of Kunming University of Science and Technology (grant number KKZ3201852050).

References

1. Tu GC, Gao ZM, Hu RZ, Zhang Q, Li CY, Zhao ZH, Zhang BG (2004) Geochemistry and ore forming mechanism of dispersed elements. Geological Publishing House, Beijing, pp 1–424 (in Chinese)
2. Zang SL (2008) Chemistry and application of rare-scattered elements. China Petrochemical Press, Beijing, pp 1–289 (in Chinese)
3. Luo YG, Sun CY, Zhang LB, Liao WX, Qu HT, Sun SJ (2015) Research on purification of unconventional elements in zinc hydrometallurgy process. World Nonferrous Metals 11:16–19
4. Zhao ZW, Ren HJ (2011) Handbook of physicochemical properties of lead zinc and its associated elements and compounds. Central South University Press, pp 1–351 (in Chinese)
5. Luo YG, Sun CY, Qu HT, Sun SJ, Ma YC, Xie TF, Wang RS, Luo LY (2015) Composite additive for purification and removal of Se and Te from zinc sulfate solution and its application, China: CN 105018729 A, 2015.11.04
6. Tugrul AB, Haciyakupoglu S, Erenturk SA, Karatepe N, Baytas AF, Altinsoy N, Baydogan N, Buyuk B, Demir E (2013) Selenium adsorption on activated carbon by using radiotracer technique. Causes, impacts and solutions to global warming. Springer, New York, pp 305–322
7. Havlik T, Dvorscikova J, Ivanova Z, Kammel R (1999) Sulphuric acid chalcopyrite leaching using ozone as oxidant. Metallurgy 53(1):57–60
8. Cotton P, Wilkinson JA (1973) Inorganic chemistry. Academia Praha
9. Gottschalk C, Libra JA, Saupe A (2000) Ozonation of water and waste water—a practical guide understanding ozone and its application. Wiley-VCH, Germany
10. Zhang WS, Cheng CY (2007) Manganese metallurgy review. Part III: Manganese control in zinc and copper electrolytes. Hydrometallurgy 89:178–188

11. Calzada LE, Gomez CO, Finch JA (2005) Nickel recovered from solution by oxidation using ozone: some physical properties. Miner Eng 18:537–543
12. Nishimura T, Umetsu Y (1992) Separation of cobalt and nickel by ozone oxidation. Hydrometallurgy 30(1–2):483–497
13. Tian QH, Guo XY, Yi Y, Li ZH (2010) Kinetics of oxidation-precipitation of cobalt (II) from solution by ozone. Trans Nonferrous Met Soc China 20:42–45
14. Nishimura T, Umetsu Y (2001) Oxidative precipitation of arsenic (III) with manganese (II) and iron (II) in dilute acidic solution by ozone. Hydrometallurgy 62:83–92
15. Lin SH (1993) Ozonation kinetics of iron and manganese in aqueous solution. J Chem Tech Biotechnol 56:163–166
16. Rischbieter E, Stein H, Schumpe A (2000) Ozone solubilities in water and aqueous salt solutions. J Chem Eng Data 45(2):338–340
17. Florinella M, Clemens VS (2000) Determination of fast ozone reaction in aqueous solution by competition kinetics. J Chem Soc Perkin Trans 2(4):661–664

Effect of Temperature on the Leachability of Chromium in EAF Slag

Ya-Jun Wang, Jun-Guo Li, Ya-Nan Zeng, and Zhi-Yuan Gao

Abstract The objective of this study is to evaluate the effect of temperature on the leachability of chromium in EAF slag. Mineral phase identification and micromorphology analysis were used to study the distribution of chromium in the EAF slag and the leaching residue. Sequential leaching tests were performed to study the chromium leachability in EAF slag under different temperature. The results show that the chromium in the EAF slag was mainly present in the Mg-Cr spinel, Fe–Cr alloy, or distributed in the form of oxide in the matrix phase-merwinite. In the sequential leaching process, the leachates under various temperatures were all alkaline and reductive. The release rates of chromium could reach their equilibrium values when the leaching time reached to 48 h, and the trivalent chromium was the predominant speciation of chromium in the leachates. The leachability of chromium in EAF slag decreased gradually with increasing temperature.

Keywords EAF slag · Temperature · Sequential leaching · Cr concentration

Introduction

Stainless-steel slag is a by-product of the stainless-steel smelting process and can be divided into EAF slag and AOD slag [1, 2]. Among them, EAF slag is produced during the preliminary stainless-steel smelting process using the EAF furnace [3]. In 2019, China's stainless-steel crude steel output reached 29.4 million tons. In most cases, for every 3 tons of stainless steel produced, 1 ton of stainless-steel slag is generated [4]. Based on this estimate, the annual output of China's stainless-steel slag is close to 10 million tons. The content of chromium in EAF slag can be as high as 5 to 15 wt% [5–7]. Due to the high toxicity of hexavalent chromium, EAF slag is

Y.-J. Wang · J.-G. Li (✉) · Y.-N. Zeng
School of Metallurgy and Energy, North China University of Science and Technology, Tangshan 063009, China
e-mail: lijg99@163.com

Z.-Y. Gao
Tangshan Research Academy of Environmental Planning, Tangshan 063000, China

B. Li et al. (eds.), *Materials Engineering—From Ideas to Practice: An EPD Symposium in Honor of Jiann-Yang Hwang*, The Minerals, Metals & Materials Series,
https://doi.org/10.1007/978-3-030-65241-8_19

usually classified as hazardous waste. In China, the recycling rate of EAF slag is less than 20%, and the amount of EAF slag in stockpiles is huge, and the best destination for this part of EAF slag is to be recycled.

The chromium in the EAF slag can migrate to the surrounding environment with the leaching action of rainwater, thereby posing a threat to the ecological environment [8]. In the water environment, chromium mainly exists in trivalent and hexavalent forms [9]. Within a certain limit, trivalent chromium is an essential nutrient element for the human body, but when people are exposed to areas with high trivalent chromium content for a long time, the human skin will become allergic and even cancerous [10]. Trivalent chromium will be converted into highly toxic hexavalent chromium under high temperature and presence of MnO_2, thereby enhancing the risk of chromium leaching [11, 12]. It can be seen that the risk of leaching chromium in EAF slag is a key factor hindering its resource reuse.

Chinese identification standards for hazardous wastes-identification for extraction toxicity (GB 5085.3–2007) stipulates that after solid waste is leached, the concentration of total chromium in the leachate must be less than 15 mg/L, and the concentration of hexavalent chromium must be less than 5 mg/L [13]. The leaching risk of chromium in EAF slag seriously restricts its recyclable value. To dispose and use EAF slag more reasonably and reduce the pollution of soil and groundwater, the migration behavior of heavy metal chromium in EAF slag should be studied. The dense urban population and high energy consumption lead to the heat island effect, making the temperature in the city higher than that in the surrounding suburbs and rural areas [14]. For areas with large temperature differences, studying the effect of temperature on chromium leachability in EAF slag can help to understand the leaching risk of chromium in different EAF slag storage sites. Qi et al. [15] studied the effect of temperature on the leaching and migration behavior of heavy metals in waste compost, and the experimental results showed that all heavy metals were easier to be leached out at higher temperatures. Alper et al. [16] studied the effect of temperature on the leachabilities of heavy metals in fly ash. The experimental results showed that when the leachate was neutral, as the temperature increased, the concentration of hexavalent chromium in the leachate increased slightly. It can be seen that temperature is one of the important factors affecting chromium leachability.

At present, the assessment of the risk of chromium leaching in EAF slag is mostly based on short-term static leaching experiments. Because mineral dissolution is limited by kinetics, the risk of chromium leaching in the long-term leaching process of EAF slag cannot be well evaluated. This study takes EAF slag as the research object and uses XRD and sequential leaching tests to study the mineral phase composition of EAF slag, the electrochemical characteristics of the leachates, and the leachability of chromium to lay the experimental foundation to clarify the influence of temperature on chromium leachability in EAF slag.

Experimental Plan

EAF Slag Sample

The EAF slag was collected from a Chinese stainless-steel manufacturer. After crushing, it was sieved through a 200-mesh (<75 μm) sample sieve. The EAF slag under the sieve less than 75 μm was placed in a drying oven and dried at a constant temperature of 105 °C for 6 h. The prepared EAF slag sample was used for physical and chemical identification and sequential leaching experiments.

Analytical Reagents

Analytical reagents $K_2Cr_2O_7$ (≥99.0%), H_3PO_4 (≥99.0%), $C_{13}H_{14}N_4O$ (≥98.0%), CH_3COCH_3 (≥99.5%), $(NH_4)_2S_2O_8$ (≥98.0%), CH_3CH_2OH (≥99.8%), $(CH_2OH)_2$ (≥99.0%), $C_{20}H_{14}O_4$ (≥98.0%), $H2SO_4$ (≥99.0%), CH_3COOH (≥99.0%), and HCl (≥99.0%) were purchased from Aladdin (Shanghai). Deionized water was obtained by a water purification system (electrical conductivity of 18.2MΩ·cm, Milli-Q Academic, Merck Millipore, Darmstadt, Germany).

Characterization

The chemical composition was determined by X-ray fluorescence (XRF, ZSX Primus II) and outputted as the oxide form after being quantified by software.

The mineralogical composition was evaluated by XRD. The XRD measurement was performed on a D/MAX2500PC automatic diffractometer to determine the mineral composition of the EAF slag. The diffraction pattern was obtained with a continuous scanning device using Cu-Kα radiation at 40 mA and 45 kV, a scanning velocity of 5° 2θ/min, and a 2θ range of 10 to 80°. Qualitative and semi-quantitative analyses were performed by the adiabatic method using JADE 6.5 software based on the PDF-2004 database [17].

Sequential Leaching Test

A 20-day sequential leaching experiment was used to study the effect of temperature on the leachability of chromium in EAF slag. 10 g dried EAF slag was weighed and transferred into a 250 ml conical flask. Then 100 mL deionized water was decanted into the conical flask. The conical flask was sealed with a lid to prevent the intrusion

Table 1 Chemical compositions of the EAF slag

Components	CaO	SiO_2	MgO	Al_2O_3	Fe_2O_3	FeO	TiO_2	MnO	Cr	V
wt.%	38.64	24.01	12.63	9.55	2.50	1.82	1.61	0.88	4.21	0.13

of carbon dioxide and oxygen in ambient air. Using a gas bath constant temperature oscillator to continuously oscillate the sealed flask for 20 days at three leaching temperatures of 20 °C, 30 °C, and 40 °C. During the leaching cycle, 80 mL of leachate was sampled from the flask every 24 h, and the same volume of deionized water was added to maintain the L/S ratio. The sampled leachates were filtered through 0.22 μm membrane filters. The electrical conductivity, pH, and redox potential (Eh) of the leachates were immediately measured. Moreover, the total chromium concentrations in these leachates were determined by using the diphenylcarbazide spectrophotography method (GB7467-87) (1995) associated with the ammonium persulfate oxidation method [18]. All of the analyses were performed in duplicate.

Results and Discussion

Properties of EAF Slag

The chemical composition of the EAF slag is listed in Table 1. The main chemical components of EAF slag were CaO, SiO_2, MgO, and Al_2O_3. The contents of CaO and SiO_2 were relatively high, with percentages of 38.64% and 24.01%, respectively. The EAF slag was alkaline, and its ternary basicity ((CaO + MgO)/SiO_2) was 2.14. In addition, EAF slag also contained a certain amount of Fe_2O_3, FeO, TiO_2, and MnO. The chromium content in the EAF slag was 4.21%, which was in the typical range.

The analysis result of the mineral composition of the EAF slag is shown in Fig. 1. XRD analysis results showed that the mineral phases present in EAF slag include merwinite ($Ca_3MgSi_2O_8$), dicalcium silicate (Ca_2SiO_4), calcium aluminosilicate ($Ca_2Al_2SiO_7$), magnetite (Fe_3O_4), chromium-bearing spinel, and eskolaite (Cr_2O_3), which are consistent with the research results of Adegoloye et al. [4] and Frías Rojas et al. [19]. It could be seen that the chromium in EAF slag mainly occurred in the eskolaite and spinel mineral phases.

pH Evolution of the Leachates

The evolution of the pH of the leachates with leaching time under three leaching temperatures was shown in Fig. 2. In the early stage of sequential leaching, the pH values of the leachates increased rapidly, reaching their maximum values on the 4th

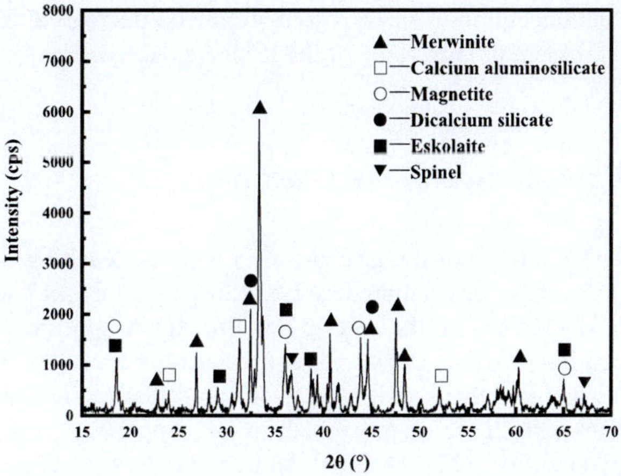

Fig. 1 XRD pattern of EAF slag

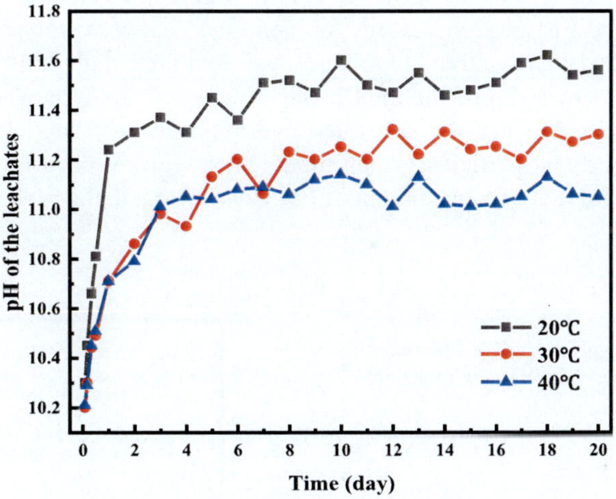

Fig. 2 Leachates' pH evolution as the function of leaching time. (Color figure online)

leaching day, and maintaining these values in the subsequent leaching time. Under neutral conditions, the dissolution rate of dicalcium silicate and magnesite is very fast, and it can reach equilibrium within 5 h [20].

When the leaching temperature was 20 °C, the pH leachate was maintained at 11.24–11.60; when the leaching temperature was 30°C and 40°C, the pH value of leachate was maintained at 10.71–11.30, showing strong alkalinity. This was because the soluble calcium silicate phases and the magnesium silicate calcium phases in the EAF slag have undergone a serious of hydrolysis reactions during the leaching process . The typical dissolution reactions were the hydrolysis of dicalcium silicate and magnesium silicate, thereby increasing the concentration of hydroxide ions in the leachate [21]. These two reactions are exothermic, so as the leaching temperature increased, the hydrolysis of dicalcium silicate and magnesia was inhibited, and the concentration of OH- ions produced decreased. Therefore, as the temperature increased, the pH value of the leachate decreased.

Eh Evolution of the Leachates

The evolution of the leachate's Eh with the leaching time is shown in Fig. 3. In the initial stage of leaching, the Eh of the leachates decreased rapidly and reached their lowest values on the second day of leaching. In the subsequent leaching stage, Eh continued to rise. During the entire leaching cycle, the Eh values of the leachates were always less than zero, which indicated that the leachates were reducing solutions. During the leaching process, the redox property of the leachate was mainly controlled by Fe^{2+}/Fe^{3+} [22]. The cristobalite in the form of divalent iron in the slag was easier to dissolve so that a higher content of Fe^{2+} ions were released into the leachate so that the leachate showed a certain degree of reduction. As the temperature increased, the Eh value increased. As mentioned above, when the temperature increased, the hydrolytic activity of minerals was inhibited, and the amount of Fe^{2+} ions leached out decreased, making the Eh value increased. As time went by, the Eh value gradually raised. Since the generated hydration product was wrapped on the surface of the reactant particles, which hindered the further occurrence of the hydration reaction and reduced the amount of Fe^{2+} ions released, the Eh value increased with time [23].

Fig. 3 Leachates' Eh evolution as the function of leaching time. (Color figure online)

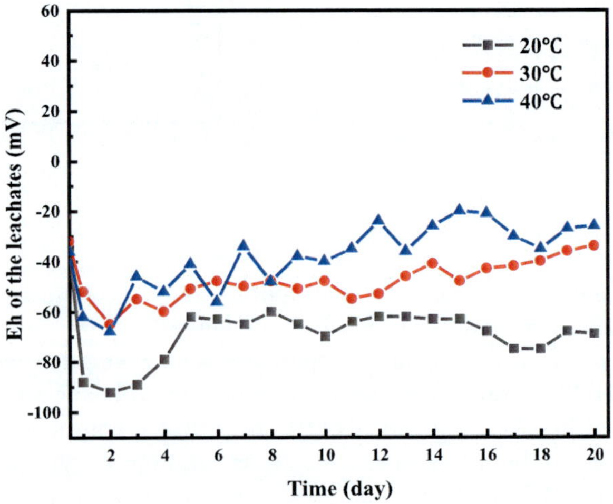

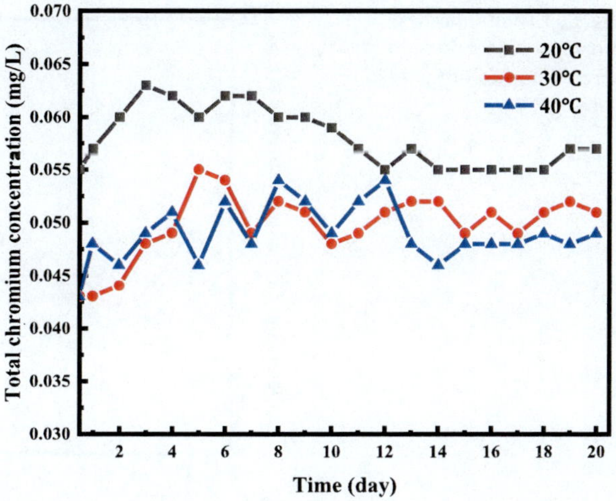

Fig. 4 Evolution of chromium concentration as the function of leaching time. (Color figure online)

Effect of Temperature on Chromium Leachability

Figure 4 shows the daily leaching amount (leaching rate mg/L/d) of total chromium at three leaching temperatures as a function of leaching time. In the early and mid-term of sequential leaching, the leaching rates of total chromium gradually increased, and in the late period leaching (after the 12th day), the leaching rates of total chromium reached their stable values. During the entire leaching cycle, the leaching rate of total chromium under 20 °C was significantly higher than 30 °C and 40 °C. The leachability of chromium in the leachate is mainly controlled by the dissolution and precipitation equilibrium of $Cr(OH)_3$ and Cr_2O_3 [24]. When the leaching temperature was 20 °C, the pH value of the leachate was higher, which promoted the dissolution of $Cr(OH)_3$ and converted $Cr(OH)_3$ to $Cr(OH)_4^-$, making the concentration of Cr^{3+} in the leachate higher [25]. In addition, when the temperature was higher, the hydrolysis reaction of the chromium-bearing mineral phases in the EAF slag was inhibited, and the concentration of Cr^{3+} in the leachate would also be reduced.

The evolution of the proportion of the cumulative leaching amount of hexavalent chromium to that of total chromium is shown in Fig. 5. During the 20-day leaching process, the leaching proportion of hexavalent chromium showed a trend of first decreasing and then increasing, and the leaching temperature had little effect on it. It could be seen that in the temperature range of this study, the change of leaching temperature was not enough to affect the existence of chromium in the leachate.

Conclusions

In this study, the method of mineral phase identification and microscopic morphology observation was used to analyze the existence of chromium in EAF slag, and then the

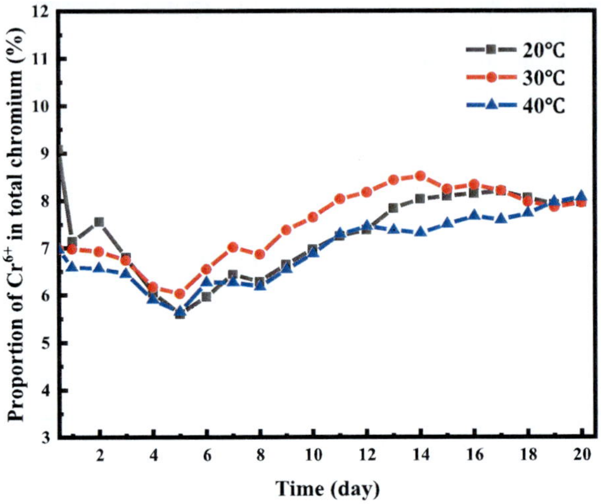

Fig. 5 Evolution of Cr^{6+} proportion in total chromium. (Color figure online)

method of sequential leaching test was performed to study the influence of temperature on the leaching characteristics of Cr in EAF slag. The main mineral phases contained in EAF slag were merwinite and dicalcium silicate. Chromium was mainly present in the EAF slag in the form of eskolaite and chromium-bearing spinel phases. The increase of the leaching temperature could inhibit the hydration reaction of the soluble mineral phases in the EAF slag, reduced the release of hydroxide ion, so that the pH of the leachate gradually decreased, while the Eh gradually increased. At lower leaching temperature, EAF slag has higher chromium leachability. The leaching temperature cannot change the valence state of chromium ions in the leachate.

Acknowledgements The authors gratefully acknowledge the support of the National Natural Science Foundation of China (No: 51704119, NO: 51574108), the Hebei Natural Science Fund Project (E2017209243), and the School Funds for Distinguished Young Scientists (JQ201714).

References

1. Liu B, Li J, Wang Z, Zeng Y, Ren Q (2019) Long-term leaching characterization and geochemical modeling of chromium released from AOD slag. Environ Sci Pollut Res 27:921–929. https://doi.org/10.1007/s11356-019-07008-7
2. Gupta T, Sachdeva SN (2020) Experimental study and modeling of concrete containing AOD steel slag for pavements. Arab J Sci Eng 45:8111–8127. https://doi.org/10.1007/s13369-020-04619-y
3. Jiang Y, Ling TC, Shi C, Pan SY (2018) Characteristics of steel slags and their use in cement and concrete–a review. Resour Conserv Recy 136:187–197. https://doi.org/10.1016/j.resconrec.2018.04.023
4. Adegoloye G, Beaucour AL, Ortola S, Noumowé A (2015) Concretes made of EAF slag and AOD slag aggregates from stainless steel process: mechanical properties and durability. Constr Build Mater 76:313–321. https://doi.org/10.1016/j.conbuildmat.2014.12.007

5. Shen H, Forssberg E, Nordström U (2004) Physicochemical and mineralogical properties of stainless steel slags oriented to metal recovery. Resour Conserv Recy 40(3):245–271. https://doi.org/10.1016/s0921-3449(03)00072-7
6. Guo M, Durinck D, Jones PT, Heylen G, Hendrickx R, Baeten R, Blanpain B, Wollants P (2007) EAF stainless steel refining–part I: observational study on chromium recovery in an eccentric bottom tapping furnace and a spout tapping furnace. Steel Res Int 78(2):117–124. https://doi.org/10.1002/srin.200705868
7. Zhao QZ, Zeng YN, Li JG, Wang YJ (2020) Selective extraction of chromium from EAF stainless steel slag by pressurized oxidation in a NaOH solution. Mater Trans 61(10):2030–2039. https://doi.org/10.2320/matertrans.MT-M2020050
8. Gao ZY, Li JG, Liu B, Wang ZM (2017) Mineralogical composition of EAF slag and its short-term leaching characteristics. Ind Safety Environ Prot 43(11):80–83. https://doi.org/10.3969/j.issn.1001-425X.2017.11.021
9. Jing C, Liu S, Korfiatis GP, Meng X (2006) Leaching behavior of Cr(III) in stabilized/solidified soil. Chemosphere 64(3):379–385. https://doi.org/10.1016/j.chemosphere.2005.12.039
10. Shadreck M, Mugadza T (2013) Chromium, an essential nutrient and pollutant: a review. Afr J Pure Appl Chem 7(9):310–317. https://doi.org/10.5897/AJPAC2013.0517
11. Eary LE, Rai D (1987) Kinetics of chromium(III) oxidation to chromium(VI) by reaction with manganese dioxide. Environ Sci Technol 21(12):1187–1193. https://doi.org/10.1021/es00165a005
12. Yang F, Guo J, Dai R, Lan Y (2014) Oxidation of Cr(III)-citrate/tartrate complexes by δ-MnO_2: production of Cr(VI) and its impact factors. Geoderma 213:10–14. https://doi.org/10.1016/j.geoderma.2013.07.022
13. State Environmental Protection Administration of China (2007) Identification standards for hazardous wastes-Identification for extraction toxicity. GB 5085.3-2007
14. Gago EJ, Roldan J, Pacheco-Torres R, Ordóñez J (2013) The city and urban heat islands: a review of strategies to mitigate adverse effects. Renew Sust Energ Rev 25:749–758. https://doi.org/10.1016/j.rser.2013.05.057
15. Qi S, Shang XJ, Zhao SL, Duo LA (2014) Proceedings of the annual conference of the Chinese society of environmental sciences. Chinese Society of Environmental Sciences and Sichuan University, pp 5936–5940
16. Baba A, Gurdal G, Sengunalp F, Ozay O (2007) Effects of leachant temperature and pH on leachability of metals from fly ash: a case study: can thermal power plant, province of Canakkale, Turkey. Environ Monit Assess 139(1–3):287–298. https://doi.org/10.1007/s10661-007-9834-8
17. Wang ZM, Li JG, Liu B, Zeng YN, Gao ZY (2017) Study on mineralogical phase composition and quantitative analysis of argon oxygen decarburization slag. Metall Anal 37(1):15–20
18. Bian LY (2007) Improvement of total chromic measure by spectrophotometry. Environ Sci Manag 32(12):146–147
19. Rojas MS, De Rojas MIS (2004) Chemical assessment of the electric arc furnace slag as construction material: expansive compounds. Cement Concrete Res 34(10):1881–1888. https://doi.org/10.1016/j.cemconres.2004.01.029
20. Engström F, Adolfsson D, Samuelsson C, Sandström Å, Björkman B (2013) A study of the solubility of pure slag minerals. Miner Eng 41:46–52. https://doi.org/10.1016/j.mineng.2012.10.004
21. De Windt L, Chaurand P, Rose J (2011) Kinetics of steel slag leaching: batch tests and modeling. Waste Manage 31(2):225–235. https://doi.org/10.1016/j.wasman.2010.05.018
22. Fällman AM (2000) Leaching of chromium and barium from steel slag in laboratory and field tests-a solubility controlled process? Waste Manage 20(2–3):149–154. https://doi.org/10.1016/s0956-053x(99)00313-x
23. Liu B, Li J, Zeng Y, Wang Z (2016) Toxicity assessment and geochemical model of chromium leaching from AOD slag. Chemosphere 144:2052–2057. https://doi.org/10.1016/j.chemosphere.2015.10.103

24. Baciocchi R, Costa G, Polettini A, Pomi R (2015) Effects of thin-film accelerated carbonation on steel slag leaching. J Hazard Mater 286:369–378. https://doi.org/10.1016/j.jhazmat.2014.12.059
25. Rai D, Sass BM, Moore DA (1987) Chromium(III) hydrolysis constants and solubility of chromium(III) hydroxide. Inorg Chem 26(3):345–349. https://doi.org/10.1021/ic00250a002

Part V
Wastewater Treatment

Development and Management of an Industrial Park for the Chinese Electroplating Industry

Yong Shi, Cuiping Huang, Zhibo Huang, Dong Xu, and Jiann-Yang Hwang

Abstract Electroplating is critical to the manufacturing but has been unwelcome to local government and residents due to its potential to cause serious environmental pollution. There are many kinds of plating, such as copper, nickel, zinc, and chromium based on the kinds of metal plated, or other varieties based on the plating methods or additives. An industrial park to house a variety of electroplating industries together has been proposed and developed. This approach provides professional operations to manage and treat the complicated environmental issues that most electroplating plants have difficulties to handle. Examples of these operations are provided.

Keywords Futianbao electroplating industrial park · Management · Safety

Introduction

With the rapid development of China's economy and technology, the center of gravity of the world's manufacturing industry has gradually shifted to China. At the same time, electroplating industry with its unique performance is becoming more and more important. Due to its strong characters, such as decorative, functionality, and versatility, it has become an indispensable industry in China's manufacturing industry.

Y. Shi · C. Huang · Z. Huang · D. Xu · J.-Y. Hwang (✉)
Futianbao Environmental Protection Technology Co., Ltd, Xian 710000, Shaanxi, China
e-mail: jhwang@mtu.edu

J.-Y. Hwang
Michigan Technological University, Houghton, MI 49931, USA

B. Li et al. (eds.), *Materials Engineering—From Ideas to Practice: An EPD Symposium in Honor of Jiann-Yang Hwang*, The Minerals, Metals & Materials Series,
https://doi.org/10.1007/978-3-030-65241-8_20

Development Status of the Industry

At present, there are more than 40,000 domestic electroplating enterprises in China. They have more than 5,000 formal production lines, with 3 billion square meters of electroplating area processing capacity [1]. These companies produce tens of billions of dollar in revenue every year. Electroplating enterprises are tightly connected to the machine manufacturing industry, light industry, electronic industry, aviation, aerospace, and instrumentation industries. Guangdong, Zhejinag, Jiangsu, and Shangdong are the four provinces has more than 75% of the total electroplating companies in China.

Enterprise Scale and Technical Level

The scale of enterprises in electroplating industry is generally small, with less than 500 enterprises having annual electroplating capacity over 100,000 m^2. A few joint ventures and formal professional enterprises have internationally advanced equipment and facilities, but most mid-small size companies still use many outdated technologies and equipment. A large number of production lines are semi-mechanized and semi-automated, and some are even manually operated.

Industry Distribution

According to statistics, 33.8% of electroplating enterprises are distributed in machine manufacturing industry, 20.2% in light industry, 5–10% in electronic industry, and the rest are mainly distributed in aviation, aerospace, and instrument industry.

Different Type of Electroplating

Galvanization is the most common type of the electroplating in China, followed by copper, nickel, and chromium plating. Among them, galvanized accounts for 45–50%, copper, nickel, and chromium plating accounts for 30%, anodized treatment accounts for 15%, electronic products, lead/tin, and gold/silver plating account for about 5% [2].

With the revitalization of China's industry and the improvement of environmental construction standards, China formulated the "Promotion of Clean Production Law of the People's Republic of China" and took effect on January 1, 2003. Therefore, through market survey, technical demonstration and reference to foreign experience, governments put forward the idea of centralized electroplating industrial park.

Through centralized management, the industrial development can be replanned, the electroplating technology can be relocated and gathered together in the centralized facility, upgraded and transformed, and the technological content of electroplating technology can be improved, so as to solve the greatest extent problem that has been plaguing the environmental protection department for a long time that the pollution sources of electroplating cannot be completely eradicated because they are too dispersed.

In recent years, the construction of industrial parks in various industries in China has made considerable progress, and a number of state-level industrial parks have been actively established, including various forms of industrial clusters, high-tech zones, economic development zones, circular economic zones, etc. At present, industrial agglomeration development with industrial parks as the main carrier has become the basic trend of industrial development in the world. Electroplating industrial parks in China have developed rapidly and become an important supporting force for economic development in many regions as well as an important carrier for the development of modern manufacturing industry in China [3].

Main Issues of the Current Electroplating Industry

Main issues of the current electroplating industry include:

1. Since there were no government agencies specially responsible for managing the electroplating industry, the distribution and development of the electroplating industry are lack of an overall and complete plan. Because factory locations are broadly distributed, scale of each is small, specialization degree is low, they cause low production efficiency and the economic benefits are poor. In recent years, with the increase of labor wages, welfare, and other factors, the increase of labor costs, rising production costs, coupled with the growth and competition of the industry, the introduction of new production equipment, large investment, greatly increase the cost of production, reduce the benefits of the enterprise.
2. Lack of complete absorption and promotion of new technologies leads to unnecessary repeated introduction and waste of funds, especially in additives and special chemicals for electroplating.
3. Due to the lack of professional training and on-the-job training, the technical level of production workers is low.
4. The level of enterprise management is low. The vast majority of electroplating enterprises still follow the extensive management mode, poor ability to adapt to market changes. Except a few international joint ventures and export-oriented enterprises, most of the enterprises do not have a sound industrial management system. Most plating enterprises are lack of liquid plating analysis and coating testing instruments and technical depth.

Table 1 Material consumption levels of several major plating species

Name	Average of advanced system (%)	Domestic average (%)
The material utilization rate of copper plating	90	65
The material utilization rate of nickel plating	90	75
The material utilization rate of chromium plating	24	10.5

Table 2 Water consumption levels of several major plating species

Best reported	Domestic advanced level	Domestic average level
0.08t/m^2 plating	0.8t/m^2 plating	3.0t/m^2 plating

5. High material consumption and low efficiency are common. The vast majority of electroplating enterprises have their material consumption and water consumption much higher than the average advanced level. According to the statistics of some enterprises that have conducted clean production audit, the material consumption and water consumption of electroplating are listed as shown in Tables 1 and 2. It is obvious that significant improvements are needed [4].

Development of Industrial Park and Solutions to the Issues

Futianbao Environment Protection Technology Ltd. is the first company in Northwest China to develop an electroplating industrial park. The company was established in 2000 and completed its first industrial park in 2003. The park has 33 acres of land and holds 40 electroplating companies (Fig. 1). Futianbao provides all the utility services and wastewater treatment service (Fig. 2). Electroplating companies can purchase or

Fig. 1 Futianbao headquarters. (Color figure online)

Fig. 2 A part of the Futianbao wastewater treatment facility. (Color figure online)

rent standardized production plants from Futianbao. Futianbao is a pioneer that had the vision for centralization of electroplating companies in China.

To fundamentally solve the problem of wastewater and waste gas pollution of electroplating enterprises in the park, it is necessary to start from the source of electroplating park design and control the whole process. The so-called standard of the whole process treatment is automatic process, overhead equipment, wastewater streams separation, and waste gas independence.

Automatic Process

Electroplating workshop production line that adopts manual frame plating processing, box basket are transferred between plating tank, and there will be a large amount of fluid and rinsing water drip. Classification and collection of each waste stream are difficult. When there are more types of production lines, the kinds of wastewater increases as well. Points of collection become more complicate. One must design innovation through technology, eliminating the park semi-automatic and manual plating production line, using automatic production line. In the design of electroplating plant, we must implement the concept of clean production, energy saving, consumption reduction, efficiency improvement, and pollution reduction by adopting advanced production facilities and equipment. According to the production scale of each enterprise, appropriately reserve the space needed to expand the scale and improve the production process, and determine the appropriate size and structure of the workshop. The size and style of the workshop should not be blindly unified, so as to avoid the restriction of the workshop and prevent the transformation of the enterprises. To consider the production of electroplating humidity, corrosive factors, choose the appropriate anti-corrosion, anti-leakage building materials, especially the ground. The ground and the corner of the joint anti-seepage measures must be strictly put in place.

Production Line Overhead Equipment

The new enterprise should be located on the second floor and above, since the production line can't be placed on the second floor. The lines shall be raised up to 50 cm above the workshop ground, to set wastewater diversion trench under. This is advantageous for the enterprise timely discovery of and repair the damaged wading pipeline, and installation of wastewater seepage system. After the electroplating park improves the production process equipment and clean production level, the wastewater treatment shall be strengthened. The sewage pipe network shall be connected to the sewage treatment station by means of open pipes and overhead, and advanced sewage treatment facilities shall be built to ensure the stable discharge of pollutants up to the standard or to achieve zero discharge.

Wastewater Diversion

Enterprises should, according to different kinds of production lines of copper, nickel, chromium, and cyanide, divide their wastewater into different streams for collection and processing. A park of electroplating wastewater treatment facility shall, in accordance with requirements of the standard, carry out the classified collection, handling, and recycling. For certain class of pollutants (total chromium, hexavalent chromium, nickel, cadmium, total silver, lead, mercury), streams must be separately collected and processed. In order to fully implement the concept of clean production, the wastewater of compatible kinds from different enterprises can be combined for treatment. A key task is to improve the reuse rate of reclaimed water after wastewater treatment and reduce the amount of freshwater consumption.

Waste Gas Independence

Based on the principle of separate collection and quality treatment, the waste gas treatment in the park should adopt the mode of independent enterprise treatment. Each company is required to install venting devices on the top and side of the production line to improve the exhaust collection rate and treatment rate. In accordance with the relevant environmental protection requirements, appropriate waste gas treatment technology for different gases will be chosen. Corresponding waste gas treatment equipment will be installed to effectively collect, treat, and discharge the waste gas emissions through the vent. Electroplating pollutants emission standards request vent shall be not less than 15 m high. Cyanide waste gas vent shall not be less than 25 m high, and all the height of the vent should be more than 5 m higher than the highest point of the building. This is beneficial to exhaust flow quickly and to ensure that the quality of atmospheric environment where the park is located.

The planning and design of the park should also pay attention to the following: risk control, online monitoring, and emergency responding system for collection system accidents, construction design, accident emergency pool, the storm water drainage collection pool, and fire control, etc. Also consideration on improving the response when excess emissions of wastewater, improving the treatment and disposal of electroplating sludge, and improving targeted and operable accident emergency response plan shall be placed. Through the whole process of risk control, the sustainable development of electroplating industry can be carried out.

Management

Electroplating park management is mainly divided into five categories, namely: environmental protection, safety, property management, wastewater treatment, and raw material management. Environmental management and safety management, as two major guarantees of electroplating park, escort the stable daily operation of the park. A modern electroplating park must first meet all the environmental and safety requirements.

In terms of environmental protection, the managers of the park will continuously publicize various environmental protection policies to various enterprises and invite the environmental protection functional departments of the government to hold special lectures for enterprises, so that enterprises can start to improve their environmental protection awareness and consciously abide by the environmental protection system. At the same time, we have formulated various emergency preventive measures. For example, the rainwater and domestic sewage pipe network of the whole park must be divided into different areas and valves should be installed in each area and the main outlet. In this way, even in daily inspection, the valves can be closed as soon as possible in case of accidental leakage, so as to avoid the expansion and outflow of the polluted water body. Each enterprise also has an independent septic tank, and each septic tank has a pH detector. This signal is uniformly transmitted to the central control room of the park. By this means, we can find out the suspected situation of stolen discharge in the enterprise at the first time. We will also conduct a 24-h water quality test and video monitoring for the industrial sewage discharge area of the enterprise, so as to ensure that the enterprise can meet the discharge standard 100% of time.

The disposal of hazardous wastes by electroplating enterprises is also a key part of the environmental protection management in the park. Because there are many kinds of hazardous wastes in these enterprises while the quantity from each enterprise is not large, it brings a lot of troubles to electroplating enterprises in handling and treatment. Our park can deal with most of the hazardous wastes of electroplating enterprises.

This not only solved a big problem for electroplating enterprises but also brought a lot of income to the park. The amount of hazardous wastes produced by each enterprise is a good reflection of the operation of the enterprise. If an enterprise maintains a saturated production amount but no hazardous wastes are produced, it means that the hazardous wastes of the enterprise must not be disposed in accordance with the regulations. Enterprises in electroplating industry need to make a lot of emergency plans and complete a lot of test reports every year. As the management operator of the park, all enterprises in the park can complete these tasks together. This can save the enterprise a lot of time, avoid a lot of red tape, and save a lot of unnecessary expenses, so that electroplating enterprises can be fully dedicated to the production. There are also several full-time environmental protection workers who inspect every enterprise in the park every day. These environmental protection workers will find problems and guide enterprises to correct them in the first time.

In terms of safety, the managers of the park will continuously publicize various safety policies to all enterprises and invite the functional departments of emergency management of the government to hold special lectures for the enterprises, so that the enterprises can start to improve their safety awareness and consciously abide by the safety system. The most common safety hazard in the electroplating industry is fire. Every quarter, we will organize electroplating enterprises to participate in various fire safety training and fire emergency drill. In terms of hardware, the campus is equipped with high-definition surveillance cameras for real-time monitoring of every corner. The park is also equipped with a professional and well-trained security team to carry out a 24-h safety inspection, escorting the safe production in the park. As for environmental protection, there is no professional security officer in the park, who conducts daily safety checks on the enterprises in the park to ensure that the enterprises meet the corresponding requirements in fire prevention, theft prevention, safe electricity use, chemicals storage, and occupational safety.

From the aspect of property, it is mainly to ensure the normal supply of water and electricity in the whole park. Because the electricity consumption of electroplating enterprises is generally large, so the electrician team of the property management should ensure that there will be no problems in the electric power of the park. Many electroplating processes need a long time of uninterrupted power consumption, and once there is a sudden power cut, it may cause great damage to the quality of products. Uninterrupted water supply is also important as the cleaning process also determines the quality of the product. Therefore, the park is equipped with professional electricians and hydraulic workers who are on duty 24 h a day to make daily inspection and regular maintenance of the public supporting system, so as to ensure the sufficient supply of water and electricity in the park. At the same time, the property management company should also be responsible for inspecting the drainage pipe network of rainwater, domestic sewage, and industrial sewage in the park. The industrial sewage pipe network needs regular inspection and maintenance.

The wastewater treatment plant, using the SCR electroplating wastewater zero discharge process independently developed by our company, is the core part of the park. The wastewater of all electroplating enterprises in the park will be discharged to the wastewater treatment plant in the park through an independent collection system and pipe network. After the system treatment, the produced water will be transported back through the pipe network to electroplating enterprises for use in the production line. The management of wastewater treatment plant is to strictly control each step of the treatment process to make the effluent meeting the standard while controlling treatment cost and equipment maintenance and use.

The management of raw materials in the park refers to the chemicals, raw materials, and various additives used by electroplating enterprises. Because only if the electroplating enterprises in the front can be controlled, then the wastewater treatment plant can be operated under stable conditions. In addition, the unified management of chemicals and raw materials in the park can improve the safety. Otherwise, each electroplating enterprise needs to set up its own independent warehouse and hire special personnel to purchase and manage these chemical agents. The unified procurement and management of the park can save many procurement costs, personnel costs, management costs, and safety risks for enterprises.

Conclusions

Electroplating industry is an integral part of China's national economy, and is an important part of industrial manufacturing, especially in precision manufacturing. Through rational policy guidance, rigorous scientific planning, strict environmental protection, and safety management, it is possible to make electroplating industry bigger and stronger. Only when the social benefit, environmental benefit, and economic benefit of the electroplating cmpanies are fully reflected, we will be welcomed by the local governments. Industrial park housing the electroplating companies together in a region is the way to go and will be promoted.

References

1. Research report on electroplating industry in China. www.cninfo360.com
2. Domestic electroplating industry status quo investigation report. 2020.5.8. https://wenku.baidu.com/view/c70ec74b0540be1e650e52ea551810a6f524c808.html#
3. Research report of China electroplating industry in 2020. 2020.7.6. https://www.qxcu.com/report/119/989473.htm
4. Ma J (2014) History of electroplating in China. Electroplating Branch of China Surface Engineering Association

Electroplating Wastewater Treatment in China

Zhibo Huang, Jiann-Yang Hwang, Cuiping Huang, and Yong Shi

Abstract Electroplating is one of the most important parts in the modern manufacturing industry. All of our modern electronics need some sort of electroplating process to help it achieve its goals. From smartphone, computer, heavy machinery, decoration, to motor vehicle, almost everything that we used nowadays need electroplating in some ways. However, at the same time electroplating factory can do much harm to the natural environment. Electroplating industry produce 2.7 billion gallons of wastewater in China every year. And most of it won't be treated to the required level and all go to the river and ocean. This causes a big issue for the environment. Dr. Hwang and Futianbao Environment Protection Technology Ltd. developed a new way to treat the electroplating wastewater. We called it "SCR", "S" means separation, "C" means concentration and "R" means recycle. The world leading edge "SCR" technology was successfully developed in 2016 by our R&D team which leads by Dr. Hwang. This technology comprehensively not only uses various physical and chemical separation technologies to achieve salt concentration and crystallization, but also recycles heavy metal from the wastewater and reuse reclaimed water. This technology meets the "zero emission" requirement in China. At the same time, it achieves the goal of harmful waste reduction and resource recycling.

Keywords SCR electroplating water treatment · Zero emission · Heavy metals

Z. Huang · J.-Y. Hwang (✉) · C. Huang · Y. Shi
Futianbao Environmental Protection Technology Co., Ltd, Xian 710000, Shaanxi, China
e-mail: jhwang@mtu.edu

J.-Y. Hwang
Michigan Technological University, Houghton, MI 49931, USA

B. Li et al. (eds.), *Materials Engineering—From Ideas to Practice: An EPD Symposium in Honor of Jiann-Yang Hwang*, The Minerals, Metals & Materials Series,
https://doi.org/10.1007/978-3-030-65241-8_21

Introduction

Electroplating is one of the most important part in the modern manufacturing industry. All of our modern electronics need some sort of electroplating process to help it achieve its goals. From smartphone, computer, heavy machinery, decoration, to motor vehicle, almost everything that we used nowadays need electroplating in some ways. Electroplating industry plays a big part in the rise of Chinese economy.

According to some study, there are more than 50 thousand electroplating companies in China [1, 2]. More than 80% of the them are in Jiangsu, Zhejiang and Guangdong provinces. Among those provinces, most of them are in Wenzhou, Yiwu, Kunshan, Suzhou, Wuxi and Chongqing (Fig. 1).

However, at the same time electroplating factory can do much harm to the natural environment. Electroplating industry produce 2.7 billion gallons of wastewater in China every year. It is about 45% of all industry wastewater being produced every year. 15 billion Chinese Yuan are being used to treat those electroplating wastewater. This is account for 1/3 of all the money being used for treating industry wastewater. From the data we can see that Chinese government and companies spend a lot of money on treating those wastewater. Even though people spend so much money on treating electroplating wastewater, more than half of it won't be able to treat to the required level and all go to the river and ocean. This causes a big issue for the environment. The water quality of electroplating wastewater is complex, it involves a

Fig. 1 The map of China, four highlight provinces has more than 75% of the total electroplating companies in China. (Color figure online)

variety of heavy metal ions and cyanide, some of them also contain carcinogenic and toxic substances, those are great harm to human. If these electroplating wastewater which doesn't meet the discharge standard and enters the environment directly, it will cause extensive and serious harm to the ecological system and human beings. The recovery of heavy metals in electroplating wastewater can let people fully use the resources. At the same time, it has great economic value. So electroplating wastewater treatment is the most important problem of industrial wastewater treatment.

Chinese government puts out a lot of policies and regulations to help and facilitate all the electroplating company to improve their technology of treating and recycling the heavy metals loaded wastewater. In section 35 of the "Cleaner Production Promotion Law of the People's Republic of China" it says: "For those companies which produces products from wastes and recover raw materials from waste, the tax authorities shall reduce or exempt those companies from value-added tax according to the relevant regulations of the People's Republic of China". So we can see Chinese government really wants company to lower the risk of environmental problems and improve the level of resources utilization. Also, Chinese government encourage company to develop better and more efficient technology to treat and recover resource from electroplating wastewater.

Conventional Treatment Technologies

There are many different kinds of process for Electroplating. For example, the zinc plating, copper plating, nickle plating, chromium plating, trivalent chromium plating, alloy plating, tin plating, etc. The technology for electroplating has advance quiet a lot in recent years. The electroplating additives have became more and more complicated, because companies wanted better finished products and more efficient electroplating process. However, at the same time it means that there are more components in the wastewater that need to be treated and recover. It makes the wastewater treatment stations face harder and harder challenges from time to time.

Right now, the common methods for treating electroplating wastewater in China include chemical method, biological method, physicochemical method and electrochemical method (Fig. 2). The chemical method includes precipitation method, oxidation method, reduction method, neutralization method and gas flotation method. Biological method includes biological flocculation method, biological adsorption method and biochemical method. Physicochemical methods include ion exchange method, membrane separation method, adsorption method and evaporation concentration method. Electrochemical methods include galvanic cell method, electrodialysis method, electroflocculation and flotation method, electrochemical oxidation method and electrochemical reduction method [3, 4].

The chemical method means that it use oxidation–reduction reactions or neutralization precipitation reaction to decompose toxic and harmful substance into non-toxic and harmless substances. Or it can use air flotation directly remove heavy metals

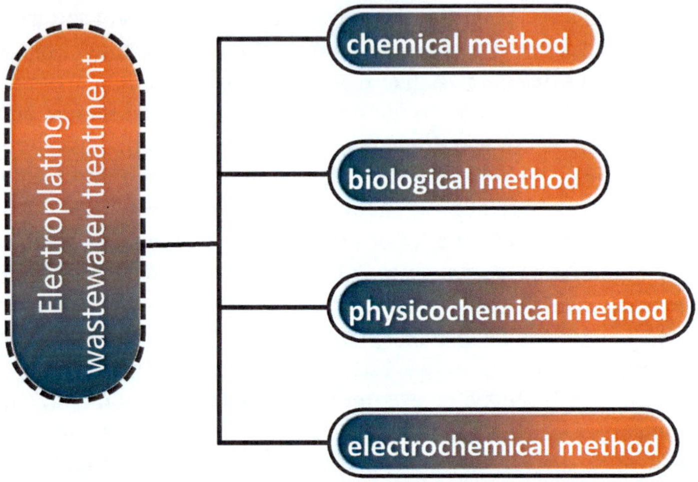

Fig. 2 The most common methods for treating electroplating wastewater. (Color figure online)

from wastewater. The chemical method is the most common way to treat wastewater. Because it requires lower initial investment, fewer machines and lower running cost compare to other methods. But at the same time, it has some drawbacks. The efficiency for the chemical method is pretty low; it can't treat the wastewater to ideal level; it can't recover heavy metals from the water, and the process produce a huge amount of electroplating sludge. However, a lot of companies still use this method to treat wastewater and this is why most of them couldn't meet the government's discharge requirement.

The study of electroplating wastewater treatment by microorganism originated in the 1980s. There are three levels of chemical, physical and genetic cooperation mechanisms. Some microbial metabolites can change the valence state of heavy metal ions in the wastewater. Meanwhile, the microbial community itself has strong biological flocculation and electrostatic adsorption, which can absorb metal ions and make heavy metals enter the sludge after solid–liquid separation, thus making the wastewater discharge up to standard. This method's advantages are its high adsorption rate, low running cost and it can recover some heavy metals from the wastewater. However, the microorganism can't survive in high concentrated wastewater or high rate of fluctuation of the components in the wastewater. Also, the microorganism has a low recycle rate. Even though some companies are still using this method, but it is mainstream anymore.

Physicochemical method use the ion exchange or membrane separation technology or adsorbents and other methods to remove impurities in electroplating wastewater, it has been widely used in industry, usually this method is used in conjunction with other methods. This method is easily managed and achieved high degree of automation. It also can treat the wastewater better compare to other methods, because it separate more heavy metals and recover them. There are some reusable water after the treatment. However, this kind of method need much more initial investment, higher running cost, require much more complicated technologies and more space.

Electrochemical treatment of electroplating wastewater began in 1940s, but it was not until 1960s that it was truly applied in the process of electroplating wastewater treatment with the development of the electric power industry. The most commonly used method is electrodialysis. Electrodialysis is a method of membrane separation. This method doesn't require additional reductant-oxidant agent in the process. And the process time is very quick. But at the same time, it consumes a lot of electricity; require higher running cost due to the fact that the reaction sometimes can be unstable and damage a lot of components.

Every method has it advantages and disadvantages. None of it can solve all the problems. Some companies can combine two methods together and achieve better result. But still most of the companies can't treat the wastewater to the require stander and hardly any company can produce reusable water from the wastewater.

China is facing worse and worse environment problems every year. Electroplating industry is one of three biggest industries which produce the highest amount of pollutants in China. At the same time we can see from the chart below that Chinese electroplating industry is growing every year (Fig. 3). If company doesn't take action and starts to find better and more efficient way to treat the wastewater, the environment would only get worse to the point that people will run out of clean water to use and drink. Our food will be contain so much toxic and harmful substance that countless

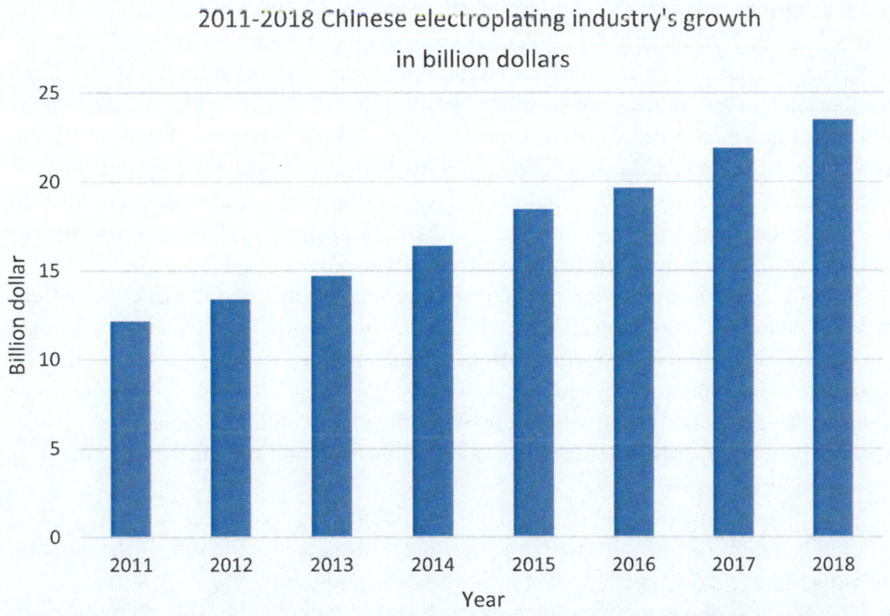

Fig. 3 Growth of Chinese electroplating industry from 2011 to 2018. (Color figure online)

people will die from all kinds of disease. Even right now there are many people dying from disease which cause by heavy metal poisoning and all the other substances that present in the electroplating wastewater. It is those electroplating companies' responsibility to solve this problem.

There are three important steps that we think can improve this situation. First, companies need to combine more methods and technologies together to treat the electroplating wastewater. Second, "zero discharge" technology is the future of wastewater treatment. It not only recycles a lot of useful resources out of the wastewater, but also save the environment from any pollution. Third, centralization of electroplating companies can then to save money on investment and operating cost of wastewater treatment station, and it also makes government easier to monitor the company to see if those companies comply with the regulations.

Futianbao Environment Protection Technology Ltd., developed a new way to treat the electroplating wastewater, as shown in Fig. 4. We called it "SCR", "S" means separation, "C" means concentration and "R" means recycle. This technology comprehensively not only uses various physical and chemical separation technologies to achieve salt concentration and crystallization, but also recycles heavy metal from the wastewater and reuse reclaimed water. This technology meets the "zero emission" requirement in China, which is essentially zero liquid discharge. At the same time, it achieves the goal of harmful waste reduction and resource recycling.

Compared with traditional processes, the water that produced by SCR system is stable and meets the requirement of Electroplating Pollutant Discharge Standard (GB21900-2008); Due to the limitation of the traditional technology itself, the system compatibility is not enough, so the water that produced by tradition method is unable to meet the standard stably.

The SCR system realizes the automatic operation of the system through a large number of sensors. In the case of water quality fluctuation, the corresponding operating parameters can be adjusted to reduce the working intensity of personnel and avoid human error. At the same time, water quality and operation parameters are automatically recorded in the background to preliminarily realize data acquisition and classification, laying a foundation for the next step of big data analysis and AI, and further optimizing the operation of the system.

Adopt the three-column ion exchange technology to ensure the effective enrichment of valuable metal ions. Through the leading electrodeposition technology, the classification and homogenization of metal ions are realized to maximize the recovery of valuable metals without secondary pollution. Compared with the traditional method of solid–liquid separation by chemical precipitation, metals are formed, no additives are needed, the recovery rate is improved and the amount of sludge is greatly reduced.

SCR process makes good use of the role of advanced oxidation, removing COD efficiently, ammonia nitrogen and other harmful substances at the same time, making the removal of heavy metals in water more easily and thoroughly.

The electrodialysis membrane separation and other concentration technology of the SCR process enable the energy saving and efficient separation of heavy metals and

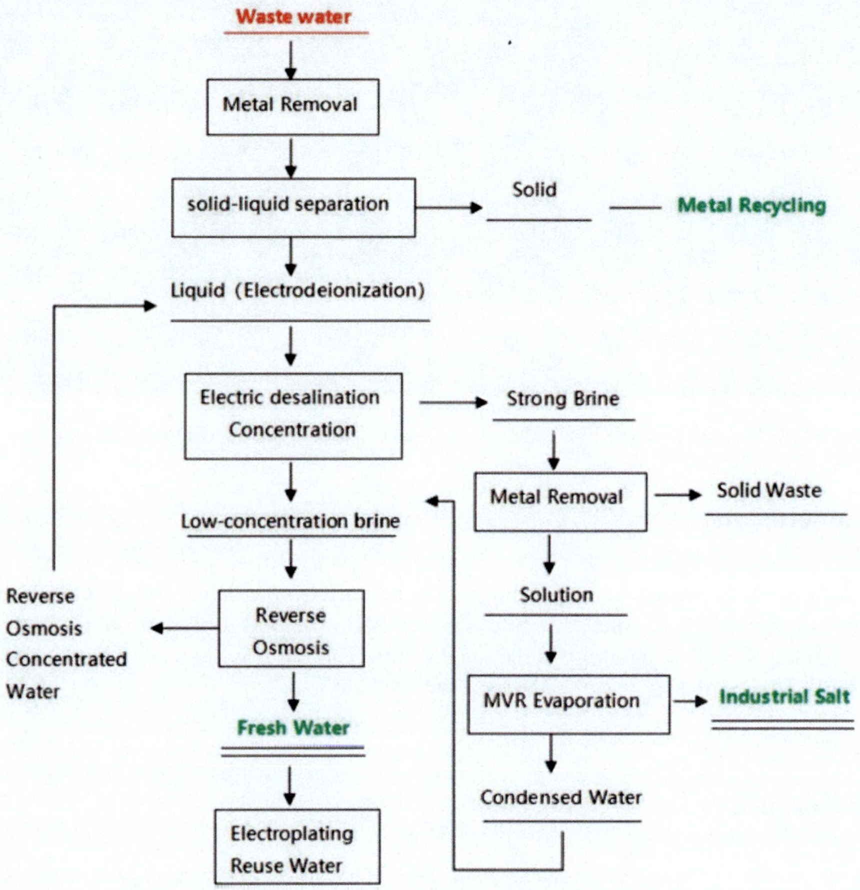

Fig. 4 Futianbao zero emission technology. (Color figure online)

concentrated salt. It is more energy saving and efficient compared with the traditional method of multi-stage reverse osmosis, and has a longer operating life.

The SCR process uses electrodialysis fresh water as the inlet water of the reverse osmosis system, which ensures the service life of the reverse osmosis membrane module.

Compared with traditional single-effect and multi-effect evaporation, The MVR evaporation adopted by SCR is more energy saving and highly automated.

The PLC control system, as shown in Fig. 5, of the whole plant can realize remote observation and control and at the same time greatly save the corresponding labor cost compared with the traditional sewage treatment plant of the same scale.

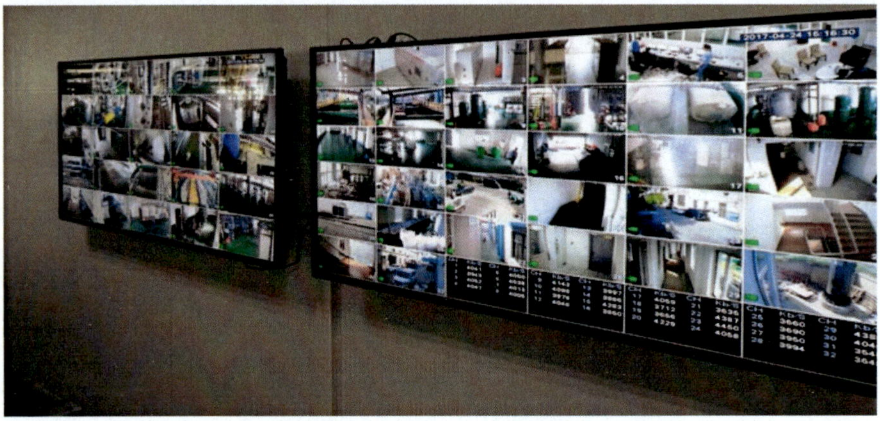

Fig. 5 PLC control system of Futianbao SCR "zero emission system". (Color figure online)

Conclusions

Electroplating wastewater is a huge problem in China. With the electroplating industry growing, people must find a way to stop the pollution that comes with it. Futianbao's SCR technology is the ideal solution so far. It will be a big step forward for the electroplating wastewater industry.

References

1. Science and technology innovation guide, 2013.03.03. https://wenku.baidu.com/view/92a020507d192279168884868762caaedc33ba14.html#
2. Current situation of electroplating wastewater treatment in China, 2020.3.31. https://wenku.baidu.com/view/ae2600d5f08583d049649b6648d7c1c708a10b87.html#
3. Analysis of basic process flow of electroplating wastewater treatment, 2018.9.13. https://wenku.baidu.com/view/b9d35fe5d0f34693daef5ef7ba0d4a7302766cdc.html
4. Duan G (2010) Electroplating wastewater treatment and reuse technical manual

A New Electroplating Wastewater Treatment Process Using Electric Lime and Vacuum Filtration

Zhixing Fu, Jiann-Yang Hwang, Yahui Sun, Yang Yang, Fujiu Nian, Lili Xi, Zhibo Huang, and Zhichao Chen

Abstract This study explores a new electroplating wastewater treatment process which could be cheaper and more efficient than the conventional approaches. The process consists of 3 steps: (1) electric lime is added to the electroplating wastewater to adjust the pH to around 9; (2) activated carbon is added to the turbid liquid; and (3) vacuum filter is used to replace traditional sedimentation tank for solid–liquid separation. Electric lime is used to replace liquid alkali for pH adjustment, which can greatly reduce production cost. Compared with the traditional sedimentation tank, the vacuum filter occupies less space, has lower operating cost and higher filtration speed, which greatly improves the production efficiency. The effluent of treated electroplating wastewater using the new process has the COD content reduced by about 60%, and the total oil content reduced by about 80%. This is obviously better than the traditional treatment process.

Introduction

Electroplating is an important basic processing industry in China. Electroplating is used to improve metal or plastic's appearance, or to increase the object's corrosion resistance [1–7]. Electroplating processes is a significant source of heavy metals such as Zn, Fe, Cr(VI), Ni, Au, and Cu in wastewater [8–12]. Washing wastewater from plated parts generally accounts for more than 85% of the total wastewater from plating plants [13]. The electroplating wastewater is harmful to the environment [14–17], as well as pose adverse effects on human health, which must be treated before being discharged [18–22].

Solid–liquid separation is an important step in electroplating wastewater treatment. Most of the traditional methods of rapid solid–liquid separation are flocculation

Z. Fu · J.-Y. Hwang (✉) · Y. Sun · Y. Yang · F. Nian · L. Xi · Z. Huang · Z. Chen
Futianbao Environmental Protection Technology Co., Ltd, Xian 710000, Shaanxi, China
e-mail: jhwang@mtu.edu

J.-Y. Hwang
Michigan Technological University, Houghton, MI 49931, USA

B. Li et al. (eds.), *Materials Engineering—From Ideas to Practice: An EPD Symposium in Honor of Jiann-Yang Hwang*, The Minerals, Metals & Materials Series,
https://doi.org/10.1007/978-3-030-65241-8_22

sedimentation, but the coagulants and flocculants used in flocculation sedimentation are new pollutants [23, 24]. This method of solid–liquid separation takes a long time and its effect is not controllable. With the improvement of domestic environmental protection requirements, we need to explore some new and more efficient methods for fast solid–liquid separation.

Rotary drum vacuum filter is a general-purpose filter equipment, mainly used in metal grinding or other processing of cooling fluid filtration, also often used in industrial water, organic solution and other solutions to remove impurities [25, 25]. The rotary drum vacuum filter relies on the gravity of the liquid itself and the vacuum negative pressure through filter cloth to isolate the sundries, so as to achieve the purpose of purifying the liquid. Its advantages are fast filtering speed, higher utilization rate of press cloth, less space occupied, lower operating cost, and better filtering effect. In this paper, we try to use a vacuum filter instead of flocculating sedimentation tank to carry out solid–liquid separation experiment of electroplating wastewater.

Experimental

Experimental reagents include the following: activated carbon (150 mesh), electric lime, polyaluminium chloride (PAC), anionic polyacrylamide (HPAM), filter cloth (200 mesh).

Calcium carbide reacts with water to form calcium hydroxide and acetylene gas, as shown by the equation below:

$$CaC_2 + 2H_2O = Ca(OH)_2 + C_2H_2 \tag{1}$$

The main component of industrial calcium carbide is calcium carbide and the rest are free calcium oxide, carbon, silicon, magnesium, iron, aluminum compounds, and a small amount of phosphates and sulfides. The remaining solid after the reaction of industrial calcium carbide with water is electric lime. The main component of electric lime is calcium hydroxide.

New electroplating wastewater treatment experiment: A moderate amount of electric lime is added to 1 ton of electroplating wastewater, adjust pH to 9, then activated carbon (200 g) is added to the turbid liquid, and stir for 30 min. Finally, the turbid electroplating wastewater liquid is pumped into a rotary drum vacuum filter for filtration.

Traditional electroplating wastewater treatment experiment: A moderate amount of sodium hydroxide solution (30%) is added to 1 ton of electroplating wastewater, adjust pH to 9, PAC solution (1000 ml, 1%) is added to the turbid liquid, and stir for 10 min. Then HPAM solution (1000 ml, 1‰) is added to the turbid liquid, and stir for 10 min. Finally, the turbid electroplating wastewater is pumped into a sedimentation tank. After 6 h in the sedimentation tank, the supernatant is discharged.

Results and Discussion

In the new electroplating wastewater treatment process (Fig. 1b), electric lime is used to replace liquid alkali for pH adjustment. Compared with traditional liquid alkali or calcium hydroxide, electric lime can greatly reduce the production cost. The addition amount of activated carbon is 0.1–1‰, and the activated carbon particle is 150 mesh.

Through experiments, we found that the COD and total oil in electroplating wastewater are the most difficult pollutants to remove, but activated carbon with good adsorption performance can significantly reduce the COD and total oil of electroplating wastewater. Wooden activated carbon, coconut shell activated carbon and coal activated carbon were tested in experiments, and a lot of activated carbon with different particle diameter were tested, including 50 mesh, 100 mesh, 150 mesh, 200 mesh, and 250 mesh. Through those experiments we found that the coal activated carbon with small particle size has better removal effect on COD and total oil. The removal rate of COD and total oil reach 50% and 80%, respectively. In addition, electric lime and activated carbon can solve the problem that the rotary drum vacuum filter is easily blocked.

The structure of the rotary drum vacuum filter is shown in Fig. 2: 1 belt conveyor, 2 scraper, 3 vacuum filter, 4 liquid material spreader, 5 reduction motor, 6 air compressor, 7 vacuum pump, 8 air water separator silencer, 9 filtrate separator.

Electric lime and activated carbon are added to electroplating waste water, and stir for 30 min. The turbid liquid is transferred to the rotary drum vacuum filter reservoir and liquid spreader (4). During filtration, the drum is immersed in the liquid storage tank and rotates slowly. The filter chamber of the vacuum filter is connected with the vacuum pump (7) (vacuum pressure 0.07–0.08 Mpa). Filtrate from the liquid storage tank and the filtrate from the liquid material sprayer are sucked out of the vacuum filter, and the solid particles are adsorbed on the filter surface to form filtrate residue.

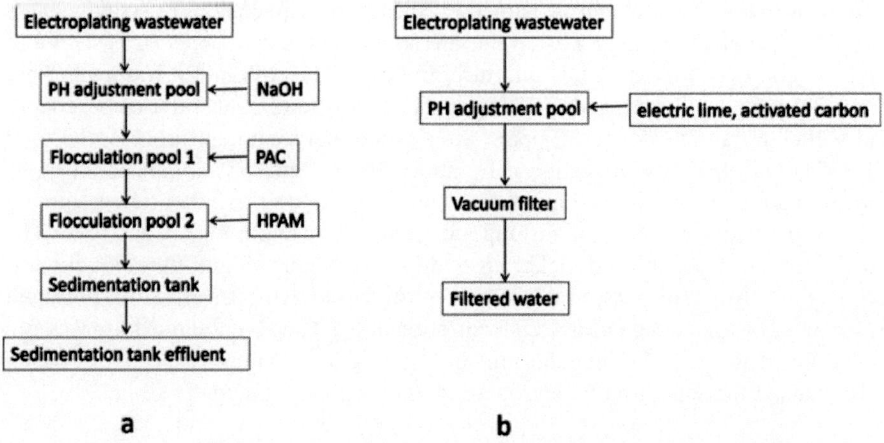

Fig. 1 Traditional electroplating wastewater treatment process (**a**) and new process (**b**)

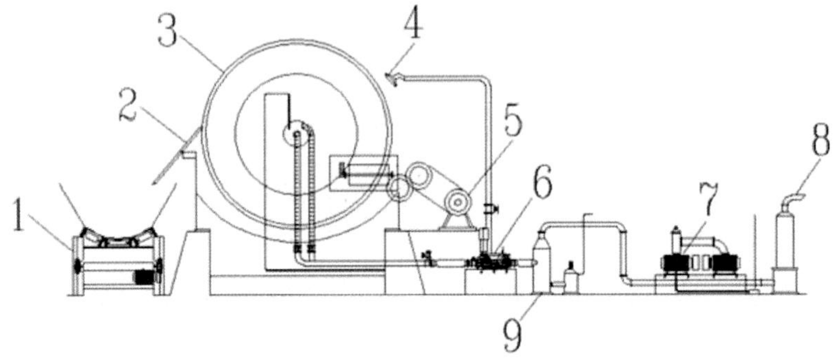

Fig. 2 The structure of rotary drum vacuum filter

When the filter chamber is rotated above the liquid level, the feed liquid is provided to the vacuum filter through the feed liquid spreader (3) to improve the filtration efficiency. With the continuous filtration, the part of the filter cake with thickness exceeding scraper (2) will be removed and transported out through belt conveyor (1). The filter chamber of the vacuum filter is connected with the air compressor (6). The air compressor backblow (0.5 Mpa) can loosen the filter cake and dredge the pores of the filter cloth.

Compared with the raw electroplating wastewater, the COD and total oil content of the water filtered by rotary drum vacuum filter are reduced by about 60% and 80%, respectively, and the COD and total oil removal rate are significantly higher than that of the traditional sedimentation tank effluent. The turbidity, copper ions, and nickel ions content of rotary drum vacuum filter effluent is also significantly lower than that of the effluent from the sedimentation tank. Table 1 shows the analysis of rotary drum vacuum filter effluent and sedimentation tank effluent.

In order to verify the reliability of the new treatment process, six experiments were carried out in this study, which are named 1, 2, 3, 4, 5, 6, respectively. The conductance and pH of the electroplating wastewater are about 5500 us/cm and 2.0, as shown in Table 1. Total oil content, COD and turbidity of the electroplating wastewater are about 6.0 ppm, 150 ppm, and 5.5 NTU, respectively, and concentration of copper and nickel are about 35 and 30 ppm. After traditional treatment process, the total oil, COD and turbidity are reduced by about 10, 11, and 57%. Concentration of copper and nickel are reduced by about 98 and 97%. However, after new treatment process, the total oil, COD and turbidity are reduced by about 81, 60, and 93%. The concentration of copper and nickel are reduced by about 99 and 99%. Obviously, the new electroplating wastewater treatment process has better effect on pollutant treatment. The total oil, COD and concentration of Cu^{2+} and Ni^{2+} of the drum vacuum filter effluent are about 1.2 ppm, 66 ppm, 0.39 ppm, and 0.28 ppm, respectively, which have reached the standard of electroplating wastewater discharge in China.

Table 1 Comparison of water quality between the vacuum filter effluent and the sedimentation tank effluent

	pH	Conductance(us/cm)	The total oil content (ppm)	COD (ppm)	Turbidity(NTU)	Cu (ppm)	Ni (ppm)
Electroplating wastewater 1	1.8	5886	6.3	155	5.10	34	26
Drum vacuum filter effluent 1	9.5	4465	1.1	68	0.34	0.39	0.25
Sedimentation tank effluent 1	9.6	4570	5.9	146	2.20	0.69	0.67
Electroplating wastewater 2	1.9	5891	6.2	159	5.21	36	24
Drum vacuum filter effluent 2	9.0	4453	1.2	65	0.33	0.33	0.24
Sedimentation tank effluent 2	9.1	4560	5.8	145	2.24	0.75	0.66
Electroplating wastewater 3	2.0	5879	6.4	162	5.26	40	35
Drum vacuum filter effluent 3	10.0	4449	1.3	66	0.34	0.42	0.31
Sedimentation tank effluent 3	10.0	4559	5.9	147	2.25	0.82	0.68
Electroplating wastewater 4	1.9	6010	6.8	183	5.41	42	30
Drum vacuum filter effluent 4	9.4	5480	1.2	68	0.33	0.43	0.34
Sedimentation tank effluent 4	9.5	5500	6.0	152	2.26	0.79	0.75
Electroplating wastewater 5	2.2	6110	7.2	178	6.11	45	32
Drum vacuum filter effluent 5	9.8	5560	1.4	72	0.42	0.39	0.28
Sedimentation tank effluent 5	9.8	5500	6.5	158	2.84	0.76	0.81
Electroplating wastewater 6	2.1	5960	6.9	166	5.83	42	28
Drum vacuum filter effluent 6	9.6	5420	1.2	61	0.37	0.38	0.26
Sedimentation tank effluent 6	9.5	5450	6.0	148	2.44	0.78	0.69

Conclusions

In this study, the rotary drum vacuum filter is used to replace the traditional sedimentation tank. Compared with the conventional solid–liquid separation method for electroplating wastewater, the new solid–liquid separation method has five main advantages. Firstly, the addition of electric lime and activated carbon solves the problem that the rotary drum vacuum filter is easily blocked. Second, the rotating drum or disc vacuum filter has a fast filtering speed, which saves the time of flocculation and precipitation (4–6 h in general), and greatly improves the production efficiency. Third, the cheap and readily available electric lime has greatly reduced production costs by replacing the more expensive liquid alkali and calcium hydroxide. Fourthly, through rotary drum vacuum filter, the treated water has less turbidity, COD and total oil content than that of the sedimentation tank. Fifthly, the new solid–liquid separation method eliminates the plate and frame filter behind the sedimentation tank and further reduces the production cost and labor cost. In summary, the rapid solid–liquid separation method of electroplating wastewater investigated in this experiment has good production efficiency and is suitable for large-scale technological application.

References

1. Ramesh Babu B, Udaya Bhanu S, Seeni Meera K (2009) Waste minimization in electroplating industries: a review. J Environ Sci Health Part C 27(3):155–177
2. Dobrevsky I, Dimova-Todorova M, Panayotova T (1997) Electroplating rinse water treatment by ion exchange. Desalination 108(1–3):277–280
3. Tan HKS (1998) A continuous stirred tank reactor study on chromic acid removal by ion exchange. Separation Sci Technol 33(8):1089–1106
4. Li ZH, Jones HR, Bowman RS, Helferich R (1999) Enhanced reduction of chromate and PCE by palletized surfactant-modified zeolite/zerovalent iron. Environ Sci Technol 33(23):4326–4330
5. Ouki SK, Kavannagh M (1999) Treatment of metals-contaminated wastewaters by use of natural zeolites. Water Sci Technol 39(10–11):115–122
6. Ahn KH, Song KG, Cha HY, Yeom IT (1999) Removal of ions in nickel electroplating rinse waters using low-pressure nanofiltration. Desalination 122:77–84
7. Orescanin V, Mikelic L, Lulic S, Nad K, Mikulic N, Rubcic M, Pavlovic G (2004) Purification of electroplating wastewaters utilizing waste by-product ferrous sulfate and wood fly ash. J Environ Sci Health 39(9):2437–2446
8. Benito Y, Ruíz ML (2002) Reverse osmosis applied to metal finishing wastewater. Desalination 142(3):229–234
9. Zhu J, Jin Q, Dongming L (2018) Investigation on two integrated membrane systems for the reuse of electroplating wastewater. Water Environ J 32(2):267–275
10. Hu QF, Wu ZY, Huang DB (2007) Treatment of electroplating wastewater by reverse osmosis (RO) process. Technol Water Treat 33(2):72–74
11. Chai X, Chen G, Yue PL, Mi Y (1997) Pilot scale membrane separation of electroplating waste water by reverse osmosis. J Membr Sci 123(2):235–242
12. Freeman HM (1988) Waste minimization audits report: case studies of minimization of cyanide waste from electroplating operation 24–62
13. Zhou YD, Sun M, Wang LJ, Xi GN (2014) Research of treatment technology for electroplating wastewater. Appl Mech Mater 487:713–716

14. Dalin Z, Xiang Li, Haibo Q (2015) Electroplating wastewater treatment and reuse project. Water Treat Technol 41(9):129–136
15. Zhongyi Z, Zhida Z (2008) Discussion on issues related to zero discharge of electroplating wastewater. Plating Finish 30(3):40–43
16. Xiang Hu, Jianfeng C, Chunxi Li (2008) Research status and prospects of electroplating wastewater treatment technology. Green Electroplating New Surf Treat Technol 12:5–10
17. Qixiang H, Yanhua H, Couyou X et al. (2010) Research status and prospects of electroplating wastewater treatment technology. Guangdong Chem Indus 37(4):128–130
18. Halliday-Bell J, Palmer K, Crane G (1997) Health and safety behaviour and compliance in electroplating workshops. Occup Med 47(4):237–240
19. El Safty AMK, Samir AM, Mekkawy MK, Fouad MM (2018) Genotoxic effects due to exposure to chromium and nickel among electroplating workers. Int J Toxicol 37(3):234–240
20. Yu-Chung T, Ya-Ching C, Po-Wen G, I-Shiang T, Jiin-Chyuan John L (2018) Characteristics of decreased pulmonary function and clinical symptoms in nickel electroplating workers. J Occup Environ Med 60(40):337–342
21. Lin SC, Tai CC, Chan CC, Wang JD (1994) Nasal septum lesions caused by chromium exposure among chromium electroplating workers. Am J Indus Med 26(2):221–228
22. Baral A, Engelken R, Stephens W, Farris J, Hannigan R (2006) Evaluation of aquatic toxicities of chromium and chromium-containing effluents in reference to chromium electroplating industries. Arch Environ Contam Toxicol 50:496–502
23. Jianyang H, Yahui S, Yang Y et al (2017) Process system for zero discharge of electroplating wastewater and resource recycling. China, Invention Patent, 201720924446.2: 2017.7.27
24. Jianyang H, Yahui S, Yang Y et al (2017) Process system and process method for zero discharge of electroplating wastewater and resource regeneration and recovery. China, Invention Patent
25. Anthony DS, Lee W, Scales PJ (2011) Models of rotary vacuum drum and disc filters for flocculated suspensions. AIChE J 57(4):951–961
26. Shao P, Darcovich K, McCracken T et al. (2015) Algae-dewatering using rotary drum vacuum filters: process modeling, simulation and techno-economics. Chem Eng J 268(15): 67–75

Effects of Ozone on COD Reduction in Electroplating Wastewater

Yahui Sun, Jiann-Yang Hwang, Lili Xi, Zhixing Fu, Fujiu Nian, Yang Yang, and Xin Chen

Abstract The electroplating industry is one of the surface treatment industries with strong versatility and a wide application range. The electroplating process mainly includes pretreatment, electroplating, and post-treatment. In order to achieve specific effects, additives such as brightener, subbrightener, surfactant, and others are indispensable in the electroplating process. These organics usually have poor biodegradability. In addition, due to the high toxicity of electroplating wastewater, biological treatment has great limitations on their decomposition. These organic compounds cause the increase of COD in electroplating wastewater, which seriously pollutes the membrane system and hinders the development of electroplating wastewater reuse process. In this study, ozone is investigated for the oxidation of organics and the results are discussed.

Keywords Ozone · Catalytic oxidation · Electroplating wastewater · Organic

Introduction

The electroplating industry is the basic industry for surface treatment of materials and has made great contributions to economic development. At the same time, electroplating is one of the three major polluting industries in the world. The annual discharge of electroplating wastewater reaches 4 billion tons in China. The wastewater may contain heavy metals such as chromium, nickel, copper, cadmium, and zinc, as well as cyanide, acid and alkali, and brighteners. Improper disposal of additives and other toxic and harmful pollutants, will cause huge pressure on the environment [1–5]. In the past five years, our company has made significant progress in the zero discharge technology of electroplating wastewater. In our research, we

Y. Sun · J.-Y. Hwang (✉) · L. Xi · Z. Fu · F. Nian · Y. Yang · X. Chen
Futianbao Environmental Protection Technology Co., Ltd, Xian 710000, Shaanxi, China
e-mail: jhwang@mtu.edu

J.-Y. Hwang
Michigan Technological University, Houghton, MI 49931, USA

B. Li et al. (eds.), *Materials Engineering—From Ideas to Practice: An EPD Symposium in Honor of Jiann-Yang Hwang*, The Minerals, Metals & Materials Series,
https://doi.org/10.1007/978-3-030-65241-8_23

found that although the content of a single species is low, there are many kinds of organic substances in electroplating wastewater , which are difficult to remove [6, 7]. The main indicator is reflected in the high COD (Chemical Oxygen Demand), which is easy to cause pollution to the back-end membrane system.

At present, flocculation and chemical oxidation are commonly used to treat electroplating wastewater COD, but conventional flocculation can only remove part of the suspended solids and flocculants, and has no obvious effect on other parts that are difficult to flocculate [8, 9]. Feton oxidation method is difficult to operate. Its treatment effect is average, and it is difficult to meet the new emission standards. As a green oxidant, ozone has a strong oxidizing ability and can undergo redox reactions directly. At the same time, the decomposition rate of ozone in water under alkaline conditions will increase, which will generate hydroxyl radicals with stronger oxidizing ability to degrade organic and inorganic substances [10, 11]. It can also increase the dissolved oxygen in the water without secondary pollution.

The preparation of ozone uses air as the raw material, which is easy to operate and has good production benefits. It is also suitable for large-scale applications [12, 13]. It should be noted that the ozone oxidation method requires frequent ventilation and back-end recycling to prevent poisoning.

This article mainly studies the effects of pH, ozone dosage, and reaction time on COD reduction for wastewater from an electroplating company. The results showed that the increase in alkalinity, the increase in ozone dosage, and the prolonged reaction time can improve the COD removal rate of electroplating wastewater.

Experimental

Electroplating Wastewater Sample

The supernatant after flocculation and sedimentation of electroplating comprehensive wastewater from an electroplating plant in Shaanxi Province is used as the treatment object. To ensure stable water quality, the same batch of wastewater is used to pretreat the supernatant. The water sample analysis results are shown in Table 1.

Table 1 Supernatant analysis results

Index	Supernatant liquid	Electroplating wastewater effluent regulation
COD, mg/L	198.0	50
NH3-N, mg/L	15.1	8
TP, mg/L	5	0.5
pH	6–11	6–9

Experimental Device and Process

The wastewater is adjusted to the corresponding pH by adding HCl or NaOH to an ozone reaction vessel made of organic glass. The ozone device uses pure oxygen as the gas source. After the oxygen comes out of the oxygen cylinder, the ozone mixture is produced by the ozone generator and is then passed into the ozone reactor. Rotameter and thermometer are used to record flow and temperature, and a designated amount of solution from the sampling port is taken at regular intervals for index analysis. After the ozone generator is turned on, it should be emptied for about 3 min. After the ozone concentration stabilizes, it is introduced into the reactor at a constant flow rate. When bubbles appear in the reactor, timing starts. The influence of pH, ozone dosage, reaction time, and temperature are the main factors investigated in this study.

Analysis Method

The analysis of all indicators follows the national standard method. Chemical oxygen demand (COD) adopts dichromate method, ammonia nitrogen adopts Nessler's reagent method, total phosphorus adopts ammonium molybdate spectrophotometry, pH adopts pH meter, temperature adopts thermometer, and 23 W ultraviolet lamp (254 nm) is used to provide ultraviolet light.

Experimental Results and Discussion

The Influence of Ozone Addition on the Treatment Effect

For each treatment, the experiments are carried out under the following conditions: water volume 1L, temperature 25 °C, pH is 10, air intake speed 60 L/h, reaction time 60 min. Ozone air intake concentration varied at 5, 20, 35, 50, 84 mg/L, with the corresponding ozone dosages of 0.3 g, 1.2 g, 2.1 g, 3 g, and 5 g, respectively.

The influence of different ozone dosages on the treatment effect is investigated. It can be seen from Fig. 1 that as the dosage of ozone increases, the removal rate of COD in wastewater also increases, showing an S-shaped curve. This is because the amount of ozone participating in the reaction increases, which promotes the oxidation reaction, thereby improving the degradation efficiency of pollutants.

At the low ozone concentrations, as the ozone concentration increases, the COD removal efficiency increases significantly, indicating that the mass transfer process of ozone at this time is the main control step of the reaction. However, when the ozone concentration is higher than 2.1 g/L, the increase in the removal rate is getting smaller and smaller. At this time, the COD removal rate increases slowly, indicating that the continued increase in the amount of ozone is subject to the solubility and contact

Fig. 1 Degradation effect of COD in water at different concentration of ozone

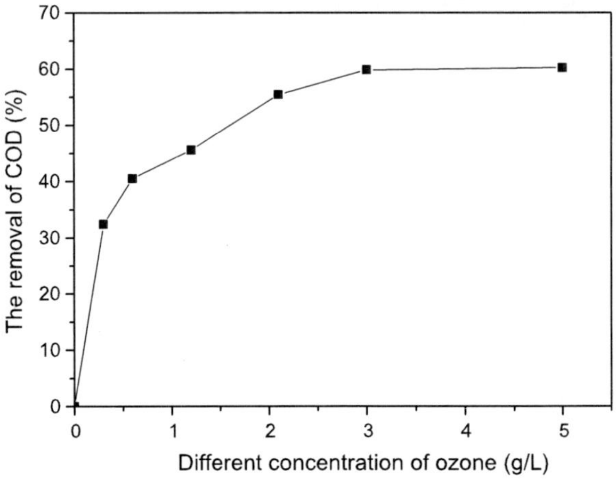

time of ozone in the wastewater, and the ozone utilization efficiency is significantly reduced. Therefore, when using ozone oxidation to treat wastewater, the intake concentration of ozone should be reasonably controlled. If the concentration is too low, the reaction time will be prolonged, and if the concentration is too high, it will cause unnecessary waste. The ozone dosage should be selected based on the concentration of the substrate. In this study, the optimal ozone dosage is chosen at 2.1 g/h.

The Influence of Wastewater Temperature on the Treatment Effect

The experiments are carried out under the following conditions: water volume 1L, pH at 10, air inlet speed 60L/h, reaction time 60min, ozone inlet concentration 35 mg/L, dosage at 2.1 g. Temperature varied from 15 °C to 35 °C and reaction time varied from 0 to 60 min. It can be seen from Fig. 2 that temperature effect on COD removal rate is relatively insignificant. At the beginning of the reaction, as the temperature increases, the COD removal rate increases slightly; but after the reaction proceeds for a period of time, the temperature increase is not conducive to the removal of COD, because as the temperature increases, the solubility of ozone in the wastewater continues to decline, weakening its transfer effect from the gas phase to the liquid phase, and the oxidation effect becomes worse. Since temperature has little effect on the treatment effect, room temperature can be selected as the temperature.

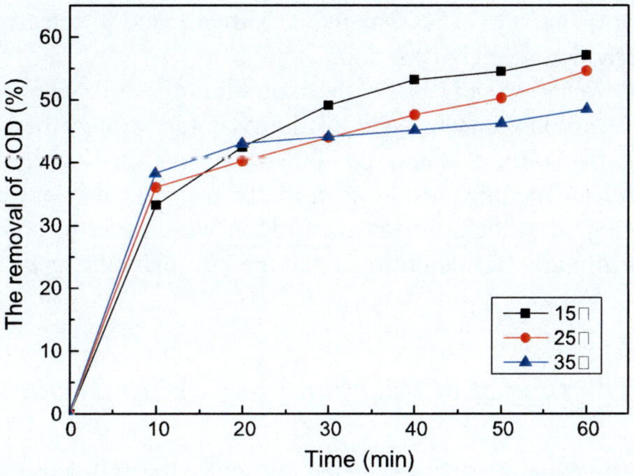

Fig. 2 Degradation effect of COD in water at different temperature. (Color figure online)

The Influence of Wastewater pH on the Treatment Effect

The experiments are carried out under the following conditions: water volume 1L, temperature 25 °C, pH is 10, air intake speed 60 L/h, reaction time 60min, ozone air inlet concentration at 35 mg/L, and ozone dosage at 2.1 g. As shown in Fig. 3, the treatment effect under alkaline conditions is better than that under acidic conditions. This is because pH affects the production of active free radicals in the reaction. The decomposition rate of ozone in wastewater increases with the increase of pH.

In the acidic region, free radicals cannot be produced at a large amount, hence the reaction is mainly a direct reaction between ozone and the reducing substances. Under neutral conditions, the reaction is manifested as direct reaction and indirect reaction at the same time. When the pH is higher than 9, a large number of hydroxide radicals in the system is produced, which is mainly manifested as an indirect reaction

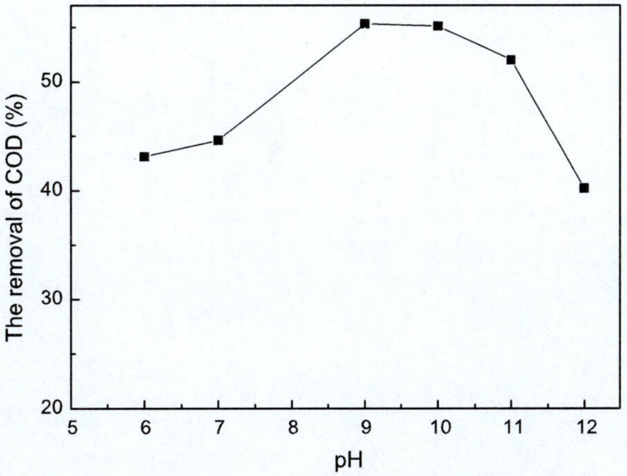

Fig. 3 Degradation efficiencies of COD in water at different initial pH

of ozone [14, 15]. Usually, the direct reaction selectivity is better, and the indirect reaction selectivity is poor.

When the pH is 11, the oxidation effect is worse than when the pH is 9. The reason is that when the alkalinity is too strong, the self-decomposition of ozone in the solution cannot be ignored, and most of the hydroxyl radicals • OH undergo self-quenching, resulting in ineffective use of ozone . Therefore, attention should be paid to the adjustment of pH in wastewater to optimize the oxidation effect. In summary, it is determined that the best pH value is 9–10.

The Impact of Reaction Time on the Treatment Effect

The experiments are carried out under the following conditions: water volume 1L, temperature 25 °C, pH 9, air intake speed 60L/h, ozone air inlet concentration at 35 mg/L, and ozone dosage at 2.1 g. As shown in Fig. 4, with the extension of the reaction time, the COD removal rate continues to increase, but the increase range continues to decrease. The reason is that as time increases, the amount of ozone added continues to increase, and the removal rate of COD in wastewater also increases. At the same time, as the COD continues to decrease, the oxidation effect becomes worse and COD removal becomes more difficult. The reaction time is as high as 150 min or more to ensure that the COD of the effluent reaches the standard. The main reasons that affect the oxidation effect are the limited solubility of ozone in the water body, the chemical reaction speed, and the increase of the reaction time. When the reaction time is 60, 150 min, COD is still as high as 90 mg/L, 51 mg/L, respectively, which is still higher than the electroplating wastewater discharge requirement (less than 50 mg/L), and there is a large amount of ozone in the tail gas that needs to be absorbed.

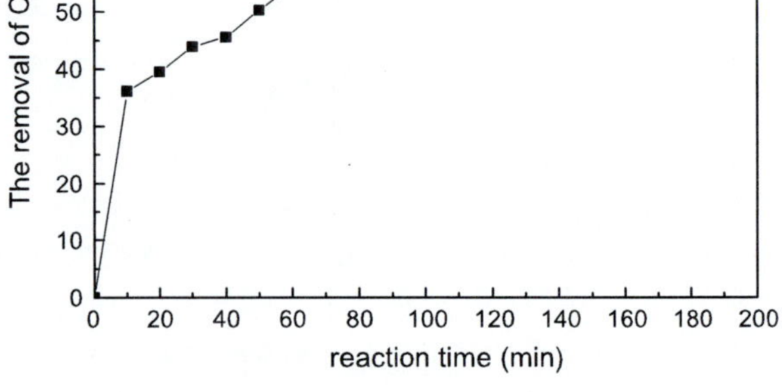

Fig. 4 Degradation efficiencies of COD in water at different reaction time

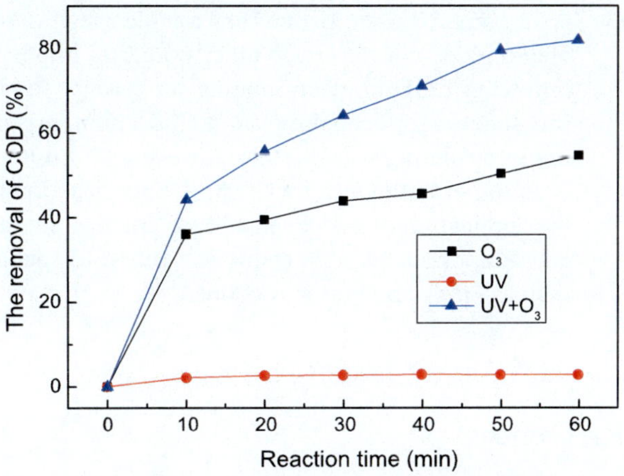

Fig. 5 Degradation efficiency of COD in water at different oxidation methods. (Color figure online)

The ozone oxidation reaction process is mainly controlled by the amount of ozone and the oxidation reaction time. To increase the COD removal effect, improve the ozone utilization rate and shorten the reaction time, it might be worthwhile to consider photocatalytic reactions [16] for hydroxyl radicals production.

The Influence of UV on Ozone Treatment Effect

Treatment conditions include: water volume 1L, temperature 25 °C, pH 9, air inlet speed 60 L/h, ozone inlet concentration 35 mg/L. Three modes are investigated: ozone only, UV only, and UV+ozone. Figure 5 shows that the UV+ozone can significantly improve the removal rate of COD, indicating high utilization rate. With the increase of time, the removal rate of COD by catalytic ozone oxidation also increases. When the reaction time reaches 50 min, the COD removal rate reaches 79.2%.

Conclusions

(1) The optimal reaction conditions are pH 10, ozone dosage at 2.1 g/L, reaction time at 60 min. The COD removal rate is 55%, and after UV irradiation, the COD removal rate reaches 80.9%.
(2) In this article, ozone alone oxidizes the COD in electroplating wastewater. The appropriate conditions are: normal temperature, pH9-10, reaction time 150 min, ozone consumption 2.1 g/(L.h) (gas flow 60L/h), which can reduce the COD of electroplating wastewater to less than 50 mg/L. However, after using ultraviolet photocatalysis, the wastewater COD can be reduced to 41.2 mg/L in about 50 min.

(3) As a green oxidant, ozone has a wide range of sources, is cheap and easy to obtain, does not introduce other polluting impurities, and has a high effect of removing COD in electroplating wastewater under weak alkaline conditions. It is suitable for deep removal of COD, ammonia nitrogen and total phosphorus in electroplating wastewater. However, the limited solubility of ozone in water is an important reason for the prolonged reaction time, which severely restricts the application of ozone. Therefore, increasing the solubility of ozone in water and adopting a suitable catalytic method to speed up the reaction will greatly increase the application of ozone.

References

1. Dalin Z, Xiang Li, Haibo Q (2015) Electroplating wastewater treatment and reuse project. Water Treat Technol 41(9):129–136
2. Zhongyi Z, Zhida Z (2008) Discussion on issues related to zero discharge of electroplating wastewater. Plat Finish 30(3):40–43
3. National Bureau of Statistics of the People's Republic of China (2016) Main pollutants in wastewater [OL]. https://data.stats.gov.cn/easyquery.htm?cn=C01
4. Xiang Hu, Jianfeng C, Chunxi Li (2008) Research status and prospects of electroplating wastewater treatment technology. Green Electroplat New Surf Treat Technol 12:5–10
5. Qixiang H, Yanhua H, Couyou X, et al (2010) Research status and prospects of electroplating wastewater treatment technology. Guangdong ChemInd 37(4):128–130
6. Jianyang H, Sun Y, Yang Y, et al (2017) Process system for zero discharge of electroplating wastewater and resource recycling. China, Invention Patent, 201720924446.2. 27 July 2017
7. Jianyang H, Yahui S, Yang Y et al (2017) Process system and process method for zero discharge of electroplating wastewater and resource regeneration and recovery. China, Invention Patent
8. Li M (2012) Research on the removal behavior of various dissolved organic pollutants in the coagulation process of the biochemical effluent of printing and dyeing wastewater. East China University of Science and Technology
9. JiaS (2016) Study on the effectiveness of two-stage MBR process in treating coal gasification wastewater biochemical effluent. Harbin Institute of Technology
10. Zhang R (2018) Research on simultaneous removal of ammonia nitrogen and COD by catalytic ozone oxidation [D]. Dalian University of Technology, 2018.
11. Tang A (2014) Experimental study on the treatment of coal chemical wastewater biochemical effluent by adsorption and ozone oxidation. Harbin Institute of Technology
12. Xi L, Sun Y, Yang Y, et al (2019) A method of removing COD from electroplating wastewater. China, Invention Patent, 201910192395.2. 14 Mar 2019
13. Jingbing L, Lei Y, Wenrong B et al (2011) Research progress of advanced oxidation technology in water treatment. Water Treat Technol 37(3):11–17
14. Brillas E, Calpe JC, Cabot PL (2003) Degradation of the herbicide 2,2-dichlorophenoxyacetic acid by ozonationcatalyzed with Fe2+andUVA light. ApplCatal B-Environ 46(2):381–391
15. Zhao L, Ma J, Sun Z et al (2008) Mechanism of influence of initial pH on the degradation of nitrobenzene in aqueous solution by ceramic honeycomb catalytic ozonation. Environ Sci Technol 42(11):4002–4007
16. Liu Q (2014) Research on advanced treatment of printing and dyeing wastewater by catalytic ozone biological activated carbon technology. Nanjing Normal University

Oxidation of Cyanide and Simultaneous Copper Electrodeposition from Electroplating Wastewater in an Electrochemical Reactor

Yang Yang, Fujiu Nian, Dong Xu, Yahui Sun, Jiann-Yang Hwang, Peiyu Qiao, Zhixing Fu, and Lili Xi

Abstract Copper was widely used in electroplating, and a lot of wastewater was produced in the rinsing process. Copper-containing electroplating wastewater can be divided into cyanide wastewater and copper sulfate wastewater according to the difference of plating process. This study is to explore the feasibility of cyanide oxidation and cupric reduction in an electrochemical reactor using an anion exchange membrane as the diaphragm. In the anolyte compartment cyanide in the cyanide wastewater was oxidized to cyanate, and furthermore, to carbon dioxide and nitrogen. In the catholyte compartment cupric in the copper sulphate wastewater was restored to copper. This study showed that the dosage of sodium hypochlorite used in the cyanide oxidation was decreased from 95 to 15 g/L and over 80% copper was recovered from copper sulphate wastewater by electrolysis.

Keywords Cyanide oxidation · Cupric reduction · Anion exchange membrane electrolyzer · Simultaneous electrolysis

Introduction

Copper cyanide plating and copper sulfate plating were widely used in copper electroplating, in which copper cyanide plating has strong adhesion and was often used as the base coating [1, 2]. Copper sulfate plating process plating, with the excellent evenness of the coating, was often used for copper coating thickening [3, 4]. Due to the difference in the composition of electroplating solution, the composition of the rinse water produced by these processes is also very different. The main components of copper cyanide wastewater are sodium cyanide, cuprous cyanide, and potassium sodium tartrate, etc., and the main components of copper sulfate wastewater are

Y. Yang · F. Nian · D. Xu · Y. Sun · J.-Y. Hwang (✉) · P. Qiao · Z. Fu · L. Xi
Futianbao Environmental Protection Technology Co., Ltd, Xian 710000, Shaanxi, China
e-mail: jhwang@mtu.edu

J.-Y. Hwang
Michigan Technological University, Houghton, MI 49931, USA

B. Li et al. (eds.), *Materials Engineering—From Ideas to Practice: An EPD Symposium in Honor of Jiann-Yang Hwang*, The Minerals, Metals & Materials Series,
https://doi.org/10.1007/978-3-030-65241-8_24

Table 1 The composition of the wastewater used in the experiment

	pH	Cu, g/L	CN, g/L	SO_4, g/L
Copper cyanide	11.0	0.5	1.8	/
Copper sulfate concentrate	1.9	25	/	44

sulfuric acid, copper sulfate, and some brightening agent. Conventionally, the treatment of copper cyanide wastewater consists of oxidation of cyanides by sodium hypochloride or chlorine and subsequent removal of copper as cupric and cuprous hydroxides [5], while the treatment of copper sulfate wastewater includes neutralization and flocculation using sodium hydroxide and lime [6], solvent extraction [7–9] and ion exchange resin separation [10–12].

The purpose of this paper is to use anion exchange membrane to divide copper cyanide wastewater and the copper sulfate wastewater, and the two wastewaters were simultaneously electrolyzed in one electrolytic cell. In the anolyte compartment cyanide was oxidized to cyanate, and furthermore, to carbon dioxide and nitrogen. In the catholyte compartment cupric in the copper sulphate wastewater was restored to copper.

Experimental

Materials and Analysis

The electroplating wastewater of the experiment was all taken from the electroplating rinsing section, and the cyanide wastewater was the rinsing water for plating parts. Copper sulfate wastewater is the concentrated solution of the rinsing wastewater treated by ion exchange. The composition of the wastewater was listed in Table 1.

The concentration of cyanide was measured by titration against a standardized silver nitrate solution as described in the standard methods. Copper content was analyzed by the spectrometric method using 1-(2-pyridine azo)-2-naphthol [13].

Experimental Procedure

According to Fig. 1, the cathode chamber and anode chamber were separated by a heterogeneous anion exchange membrane, and the effective volume of each chamber is 4L. Anodes used in the experiment were vertical 12 cm * 25 cm plates, made of Ti-based PbO_2, while cathodes were made of copper mesh with the same size. The distance between the cathode and the anode is 6 cm. The experiment was operated in batch modes. For bath tests, 4L of the wastewater containing a known concentration of cyanide or copper was poured into the reactor equipped with the previously described

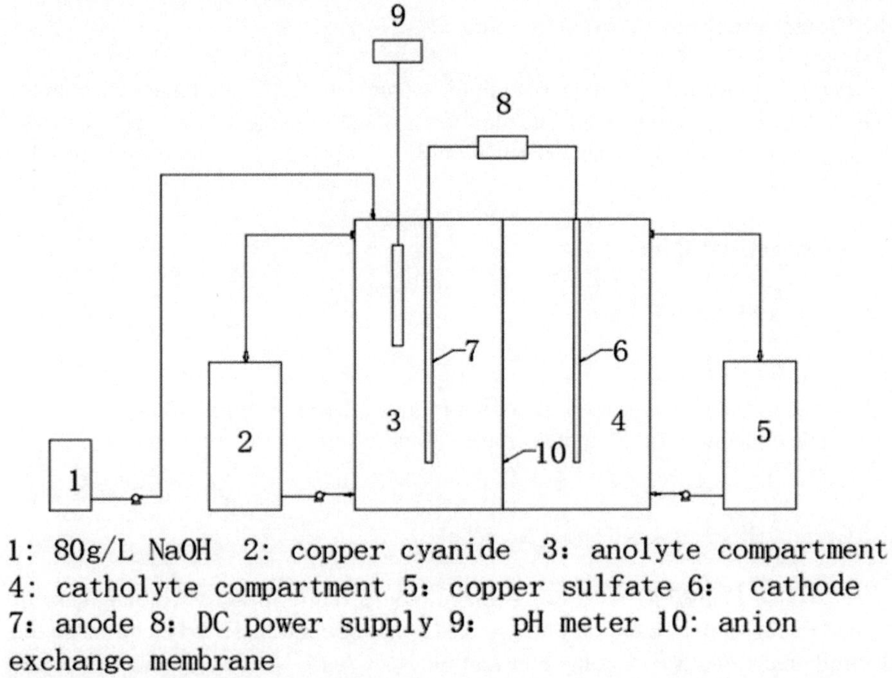

Fig. 1 A schematic representation of the experimental setup

electrodes. The density of the electrical current was regulated at the desired value, the electrical current was established between electrodes, and the reaction was started and allowed to continue up to the specified time. During the experiment, the pH of the cyanide-laden wastewater was maintained by sodium hydroxide. All experiments were conducted in the temperature range 20–30 °C. The effects of current density, residence time, and the pH of the cyanide-laden wastewater on the oxidation of cyanide and copper recovery were investigated.

The Analysis Indicators

The Removal of Cyanide

The Removal Rate of Cyanide

$R_1 = (C_0 - C_t)/C_0 * 100\%$

C_0: The initial concentration of cyanide in the anodic solution, mol/L.

C_t: The concentration of cyanide in the anodic solution in t time, mol/L.

The Current Efficiency of Cyanide Removal

$\eta_1 = V(C_0 - C_t)nF/ItM * 100\%$; I: magnitude of current, A; M: molar mass of cyanide, 26 g/mol; t: retention time, s; n: The number of transferred electrons in the electrode reaction, 5; F: faraday constant, 96485 C/mol; V: the volume of anodic solution, L.

The Recovery of Copper

The Recovery Rate of Copper

$R_2 = (C_0 - C_t)/ C_0 * 100\%$

C_0: The initial concentration of copper in the cathodic solution, mol/L.

C_t: The concentration of copper in the cathodic solution in t time, mol/L.

The Current Efficiency of Copper Removal

$\eta_2 = V(C_0 - C_t)nF/ItM * 100\%$; I: magnitude of current, A; M: molar mass of copper, 63.5 g/mol; t: retention time, s; n: The number of transferred electrons in the electrode reaction, 2; F: faraday constant, 96485 C/mol; V: the volume of cathodic solution, L.

The reactions occurring in the electrolytic cell can be written as follows:

Anodic reactions:

$2CN^- + 8OH^- - 10e^- = 2CO_2 + N_2 + 4H_2O.$

$CN^- + 2OH^- - 2e^- = CNO^- + H2O.$

$4OH^- - 4e^- = 2H_2O + O_2.$

$Cu^+ - e^- = Cu^{2+}$

Catholic reactions:

$Cu^+ + 2e^- = Cu.$

$2H^+ + 2e^- = H2.$

Results and Discussion

Effect of Current Density on Cyanide Removal and Copper Recovery

The effect of current density on cyanide removal and copper recovery was studied under the conditions of maintaining the pH value of the anodic solution being 11.0. The residence time of the anodic solution was 6 h, and the residence time of the cathode solution was 12 h. The results were shown in Figs. 2 and 3. As can be seen from Fig. 2, the cyanide removal increases with the increase of current density, while

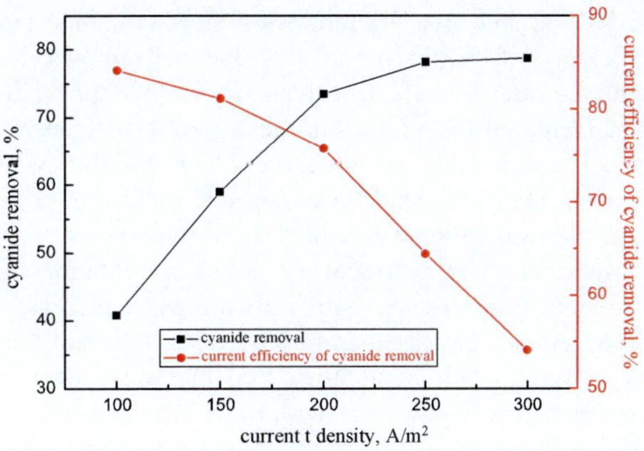

Fig. 2 Effect of current density on cyanide removal and its current efficiency. (Color figure online)

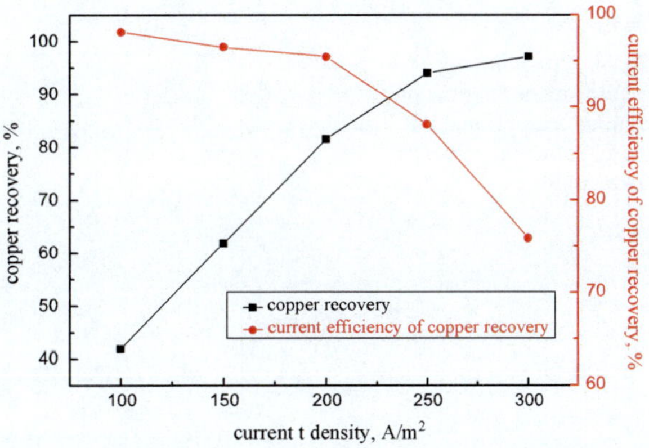

Fig. 3 Effect of current density on copper recovery and its current efficiency. (Color figure online)

the current efficiency shows an opposite trend. The copper recovery shows a similar regularity with cyanide removal. The reason for this phenomenon is that the increase of current density will lead to the increase of side reactions, which will affect current efficiency and further affect energy consumption. Therefore, the current density is set at 200 A/m^2.

Effect of Retention Time on Cyanide Removal and Copper Recovery

Figures 4 and 5 show the experimental results under different residence times. The experimental conditions are as follows: the current density of anode and cathode is 200 A/m^2, and the pH value of the anodic solution maintained in the electrolysis process is 11.0. As can be seen from Fig. 4, when the residence time of cyanide wastewater increased from 4 to 6 h, the removal rate of cyanide increased from 56.6

to 73.3%, and then the removal almost remained constant after 6 h. The current efficiency of cyanide removal shows the opposite trend, and decreases gradually with the extension of electrolysis time. When the residence time of cyanide wastewater increased from 6 to 8 h, the current efficiency decreased from 75.7 to 57.1%. The reason for this phenomenon is that with the extension of electrolysis time, the concentration of cyanide in anodic solution decrease, and the proportion of side reaction increases. Figure 5 indicates the influence of residence time of copper sulfate wastewater on copper recovery and its current efficiency. As shown in Fig. 5, when residence time of copper sulfate wastewater increased from 10 to 13 h, the copper recovery rate increased from 69.1 to 86.1%, and the residence time continued to be extended, with no significant change in recovery rate. The current efficiency of copper recovery decreases with time, but the overall change is not obvious. Therefore, in the subsequent experiment, the retention time of the cyanide wastewater was 6 h, while that of the copper sulfate was 13 h.

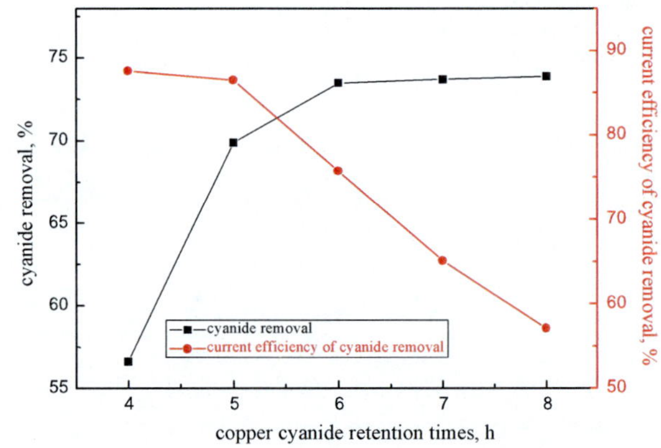

Fig. 4 Effect of copper cyanide retention time on cyanide removal and its current efficiency. (Color figure online)

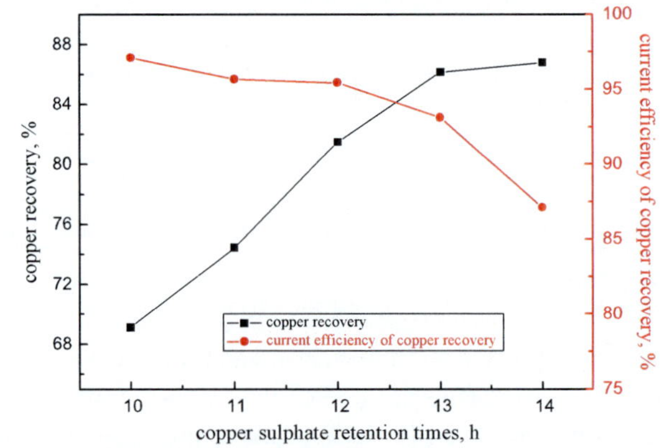

Fig. 5 Effect of copper sulfate retention time on copper recovery and its current efficiency. (Color figure online)

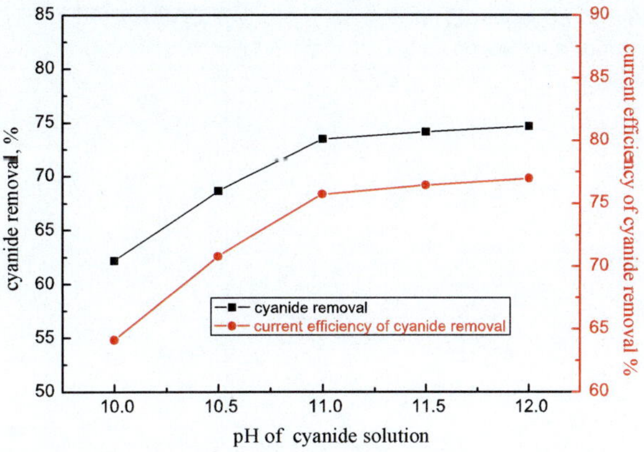

Fig. 6 Effect of pH value of cyanide solution on cyanide removal and its current efficiency. (Color figure online)

Effect of Cyanide Solution pH on Cyanide Removal and Copper Recovery

Figure 6 shows the effect of the pH of the anodic solution on the oxidation of cyanide and its current efficiency. The experimental conditions were as follows: the current density was controlled at 200 A/m^2, and the residence time of anodic solution and cathode solution was set to 6 and 13 h. As can be seen from the Fig. 6, when the pH value of the anodic liquid increased from 10.0 to 11.0, the cyanide removal rate and its current efficiency increased to a certain extent, but the effect was not obvious when the pH value continued to increase. It was also found that the pH of the anodic solution also affected the cell voltage. Figure 7 shows the cell voltage at different pH values. With the increase of pH, the conductivity of the solution gets better and better, and the cell voltage gets lower and lower. The pH value of the anodic solution has only an effect on the oxidation of cyanide and almost no effect on the recovery of copper. However, when the pH value is greater than 12.0, blue copper hydroxide precipitation is easy to occur on the ion exchange membrane (anode side), which affects the electrical conductivity of the membrane. In view of this, the pH value of the anodic solution should be controlled between 11.0 and 11.5.

Conclusion

Experiments confirmed that the feasibility of cyanide oxidation and cupric reduction in an electrochemical reactor using an anion exchange membrane as the diaphragm. The process of simultaneous oxidation of cyanides and electrodeposition of copper from electroplating wastewater proved to be feasible if conducted at a current density of 200 A/m^2 and a proper cyanide solution pH value of 11.0–11.5. When the residence time of the cathode solution and anodic solution is 13 h and 6 h, respectively, it was

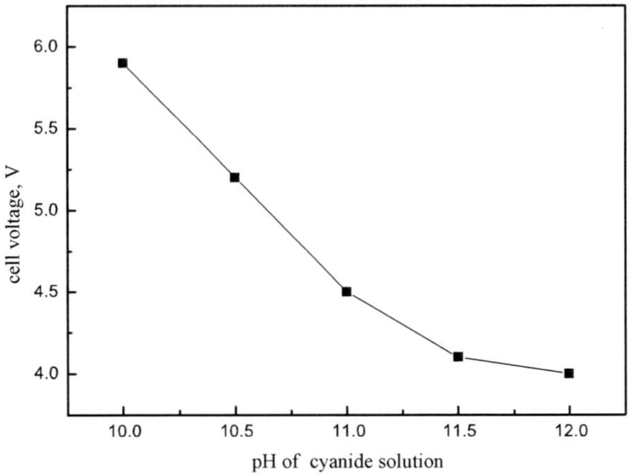

Fig. 7 Effect of pH of cyanide solution on cell voltage

possible to destroy nearly 75% of cyanides and recover 80% of copper from the solution. Therefore, it can be concluded that the co-electrolysis with an anion exchange membrane is an efficient, reliable, and promising technology for the treatment of copper cyanide and copper sulfate wastewater.

References

1. Horner J (2008) Cyanide copper plating. Plat Surf Finish 95(11):15–17
2. Davidson T (2013) Alkaline copper-plating process eliminates cyanide and improves product quality. ChemInform
3. King RD, Montgomery ER (1992) High-throw acid copper plating using inert electrolyte
4. Tench D (1985) Cyclic pulse voltammetric stripping analysis of acid copper plating baths. J Electrochem Soc 132(4):831–834
5. Xiaojun R, Yanfeng Li, Guanghui Z et al (2009) Progress in the study of the treatment of cyanide-containing wastewater from industries. Ind Water Treat 08:7–11
6. Li Y, Zeng X, Liu Y et al (2003) Study on the treatment of copper-electroplating wastewater by chemical trapping and flocculation. Sep Purif Technol 31(1):91–95
7. Li L, Zhong H, Cao Z et al (2011) Recovery of copper (II) and nickel (II) from plating wastewater by solvent extraction. Chin J Chem Eng 19(6):926–930
8. Li LQ, Zhong H (2011) Separation and recovery of copper (II), nickel (II) from simulated plating wastewater by solvent extraction using Lix984. Adv Mater Res 365:252–259
9. Separation and recovery of copper (II), nickel (II) from simulated plating wastewater by solvent extraction using Lix984. In: Proceedings of 2010 first international conference on cellular,molecular biology, biophysics and bioengineering, vol 3 (2010)
10. Choi JW, Song MH, BediakoJK et al (2020) Sequential recovery of gold and copper from bioleached wastewater using ion exchange resins. Environ Pollut 115–167
11. Gaikwad RW, Sapkal RS, Sapkal VS (2010) Removal of copper ions from acid mine drainage wastewater using ion exchange technique: factorial design analysis. J Water Res Prot 1(2):984–990
12. Siu PCC, Koong LF, Saleem J et al (2016) Equilibrium and kinetics of copper ions removal from wastewater by ion exchange. Chin J Chem Eng 24(001):94–100
13. APHA (2005) Standard methods for the examination of water and wastewater 21th ed

Extraction of Cerium from Catalyst of Waste Automobile Exhaust Gas Purifier

Chen Ailiang, Luo Guanwen, Mao Jiale, Pan Yujun, Lu Sujun, Ma Yutian, Chen Shengli, Du Zuojuan, Bowen Li, and Qiao Jinxi

Abstact Automobile exhaust catalyst has been more and more widely used for creating a green life. It contains abundant platinum, palladium, rhodium, and rare earth cerium, compared with primary resources. It's necessary to efficiently recover valuable metals with great application value. The feasibility of leaching cerium from hydrochloric acid system is mainly analyzed from the perspective of thermodynamics in this paper. The leaching experiment of hydrochloric acid oxidation is carried out. The results show that the leaching rate of cerium reaches only 46% when the concentration of hydrochloric acid was 4 mol/L, the reaction temperature was 80 °C, the amount of hydrogen peroxide was 10 ml, the liquid–solid ratio was 10:1, the reaction time was 2 h, and the stirring speed was 450 r/min. According to the metal equilibrium, it is inferred that the formation of stable Ce–Zr solid solution makes the lower leaching rate of cerium.

Keywords Rare earth · Cerium · Leaching process · Secondary resources

Introduction

Rare earth is a strategic metal element. Countries all over the world regard rare earth materials as important materials for economic development and social progress. It has been reported that [1–4] the recovery of rare earth elements from products only accounts for 1% of the recovery of rare earth elements. Meanwhile, only 10% of the global rare earth elements can be recovered and utilized, while the remaining rare

C. Ailiang · L. Guanwen · M. Jiale · P. Yujun · D. Zuojuan · Q. Jinxi (✉)
School of Metallurgy and Environent, Central South University, Changsha 410083, China
e-mail: Qiaojinxi@csu.edu.cn

L. Sujun · M. Yutian · C. Shengli
State Key Laboratory of Nickel and Cobalt Resources Comprehensive Utilization, Jinchang 737100, China

B. Li
Michigan Technological University, Houghton, MI 49931, USA

B. Li et al. (eds.), *Materials Engineering—From Ideas to Practice: An EPD Symposium in Honor of Jiann-Yang Hwang*, The Minerals, Metals & Materials Series,
https://doi.org/10.1007/978-3-030-65241-8_25

earth elements are not fully utilized. In order to realize sustainable development of resources, it is urgent to recover cerium from secondary resources.

In order to alleviate the contradiction between supply and demand and environmental pressure, countries around the world have taken certain measures to recover rare earth secondary resources: Germany and Japan have focused on waste electronic equipment. In addition, Germany also recycles rare earths from waste magnetic materials and waste automotive catalysts (abbreviated as waste catalysts) [5]. At present, the main secondary resources of cerium are waste phosphor, molten salt electrolytic slag, waste automobile exhaust gas purifier catalyst and waste petroleum catalytic/cracking catalyst and so on. According to statistics, the global output of rare earth catalysts accounts for about 20% of the total output of rare earth products, and the proportion will maintain a continuous growth trend [6]. The proportion of rare earth elements contained in these wastes is higher than that in rare earth ores [7].

Studies [8] have shown that inorganic acids such as hydrochloric acid, sulfuric acid nitric acid, and so on have a better dissolution effect on rare earth elements than organic acids. It is easy for platinum, palladium, rhodium, and rare earth elements to form coordination ions with chloride ions in aqueous solution. Many researchers choose to use chloride ion system to leach platinum, palladium, rhodium, and rare earth elements. For example, Zheng [9] used spent petroleum catalytic cracking catalyst as raw material to study the leaching rate of rare earth elements in hydrochloric acid, sulfuric acid, and nitric acid systems. The experimental results showed that hydrochloric acid and nitric acid have high leaching rate which could reach more than 90%. In addition, Qin [10] treated waste petroleum catalytic/cracking catalysts with inorganic acids. The leaching rate of rare earth reached 98%. But Wu [11] leached waste automobile exhaust catalysts with inorganic acids, and the leaching rate of rare earth was 50%. It was not thoroughly studied and discussed the reason that the leaching rate was only 50%. At the same time, the rare earth leaching rates of the exhaust gas catalysts differ greatly from those of catalytic/cracking catalysts. Thus, in order to elucidate the leaching mechanism of cerium from exhaust catalysts of automobiles, the thermodynamic analysis of Ce–Cl–H_2O system and the experimental study on the leaching of cerium from hydrochloric acid system are specially carried out in this paper.

In this experiment, the waste catalysts of a Chinese company was used as the experimental raw material. The leaching method was used to carry out the following experimental studies:

(1) Thermodynamic analysis. (2) Recovery of rare earth elements in the spent automobile catalysts in hydrochloric acid system, and single factor condition experiments on hydrochloric acid concentration, reaction temperature, amount of hydrogen peroxide, reaction time of liquid–solid ratio were carried out to determine the optimal experimental conditions, and the leaching mechanism of cerium in spent automobile exhaust catalysts was ascertained.

Thermodynamics

Ce–Cl–H_2O System

The E-pH diagram of cerium in Ce–Cl–H_2O system was drawn by HSC Chemistry 6.0. The different reaction conditions were analyzed.

Figure 1 shows the E-pH diagrams at different temperatures that can be obtained assuming that the activity of ions in the solution is 1 according to Gibbs free energy Δ_f G of different components at different temperatures.

It can be seen from Fig. 1 that cerium exists as Ce^{4+} when the potential is higher than 1.8 V and the pH is less than −2. With the increase of pH, oxides of cerium are gradually formed. As the potential decreases, cerium exists in the form of III valence or even as a single substance. Due to the occurrence of cerium and Cl^- complexation reaction, there can be a variety of complexes of cerium and chlorine in the thermodynamic stable region of the system water. With the increase of pH value, $CeCl^{2+}$ and $CeCl_2{}^+$ are generated in turn. When the pH value continues to increase to 5, the low-valent oxide of cerium CeO^+ appears. Stable CeO_2 will be formed under high pH and high potential conditions.

When the temperature increases, the stability zones of Ce^{3+} and Ce^{4+} decrease. In contrast, the stability zones of CeO_2 gradually increase. While the stability zones of the complex move toward the direction of low potential and low pH value with the increase of temperature, the stable existence area of Ce^{3+} and complex shrinks while the stable area of CeO_2 increases. Dissolution of oxides requires higher acidity and potential conditions. High concentration of acid and substances with reduction potential less than CeO_2 need to be added to reduce CeO_2 to Ce^{3+}.

According to the principle of simultaneous equilibrium and charge balance, the total content of cerium and chlorine is constant throughout the system. In order to

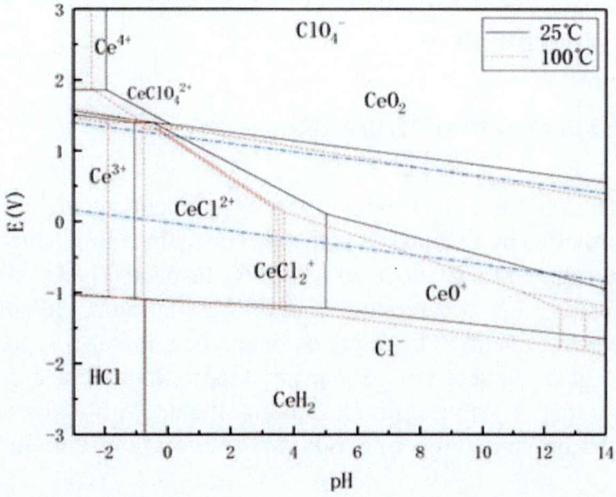

Fig. 1 E-pH diagram of Ce–Cl^-–H_2O system. (Color figure online)

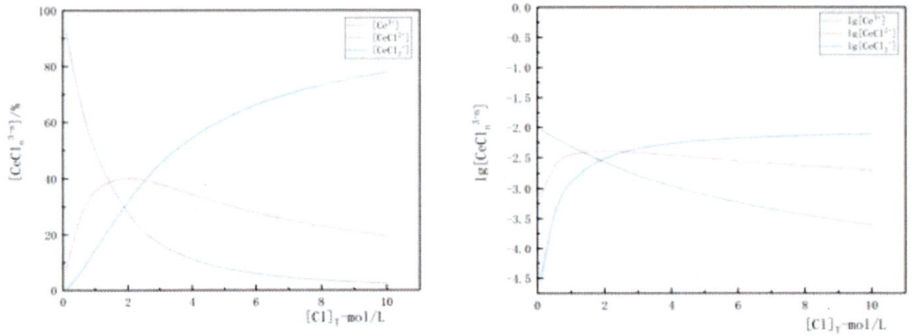

$[CeCl_n^{3-n}]\%$-$[Cl]_T$(a) lg$[CeCl_n^{3-n}]$-$[Cl]_T$(b)

Fig. 2 $[CeCl_n{}^{3-n}]\%$-$[Cl]_T$ (**a**) lg$[CeCl_n{}^{3-n}]$-$[Cl]_T$ (**b**). (Color figure online)

simplify the calculation, the activity is expressed by concentration and the reaction is approximated to equilibrium state. According to the obtained equilibrium constant of complexation reaction and other data, the relationship diagram of $[CeCl_n{}^{3-n}]$-$[Cl]_T$ is drawn as shown in Fig. 2.

Ce^{3+}, $CeCl^{2+}$, and $CeCl_2{}^+$ are the main complexes in Ce–Cl–H_2O system. With the increase of chloride ion concentration, the concentration of Ce^{3+} decreases from 92.37 to 2.47%. The concentration of $CeCl^{2+}$ reaches its maximum at $[Cl]_T = 2$ mol/L, and then decreases. The concentration of $CeCl_2{}^+$ gradually increases and finally reaches at 77.59%. It can be seen that Ce^{3+} dominates when the chloride ion concentration is less than 1.75 mol/L and $CeCl_2{}^+$ dominates when the chloride ion concentration is higher than 2.2 mol/L. Combined with the E-pH diagram of Ce–Cl–H_2O system, lower Cl^- concentration can be maintained during the leaching of rare earth elements to ensure the high efficiency of cerium leaching.

Experiment

Experimental Materials

The raw materials used in the experiment were catalysts for exhaust gas purifier provided by a Chinese company. The main components were Al_2O_3 and SiO_2, which accounted for 39.06% and 31.66%, respectively. In addition, it also contains 3.91% cerium, 3.83% zirconium and a little platinum, palladium, rhodium, and other rare earth elements. The relative content of mineral components is shown in Table 1.

It can be seen from Table 1 that the main mineral composition of the waste catalyst sample is magnesium aluminum silicate with a content of 71.15%. Among them, cerium-containing minerals are alumina and zirconium-cerium mixture, alumina

Table 1 Statistical results of relative contents of mineral components in spent catalysts (%)

Substance	Magnesium aluminum silicate	Biotite	Feldspar	Calcite	Diopside	
Content	71.15	3.11	0.53	0.15	0.03	
Substance	Quartz	Manganese spinel	Chromite	Chromite-Pb	Chromite-MnNi	
Content	0.12	0.32	0.09	0.20	0.07	
Substance	Magnetite	Manganese iron ore	Lead oxide	Manganese carbonate	Zinc apatite	
Content	0.69	0.11	0.16	0.29	0.05	
Substance	Alumina	Alumina + zirconium cerium 1	Alumina + lead cerium	Alumina + zirconium cerium 2	Alumina titanium	
Content	1.54	15.60	0.37	1.29	0.23	
Substance	Cerium Zirconium Mixture	Cerium Zirconium Oxide	Zirconia	Cerium Zirconium Oxide	CeO	Total
Content	0.42	0.37	0.36	2.54	0.18	100

and lead-cerium mixture, zirconium-cerium mixture, zirconia cerium and CeO. It occupies 20.77% of the total.

The reagents used in the experiment were concentrated hydrochloric acid and hydrogen peroxide. All of which are analytical pure. Pure water is made in the laboratory.

Experimental Methods

10 g waste catalyst is added into a beaker containing diluted hydrochloric acid. Put the beaker into a thermostatic water bath for magnetic stirring and set the corresponding speed at the appropriate temperature. After the reaction, vacuum pump was used to extract the leacheate. ICP-AES were used to analyze the concentration of related elements in the leacheate. According to the following formula:

$$\eta = \frac{CV}{m\omega}$$

where η is the leaching rate of elements, C (mol/L) is the concentration of elements in the leachable liquid, V (ml) is the volume of leachable liquid, M (g) is the quality of experimental raw materials, and ω is the element content in raw materials (wt%).

Results and Discussion

Effect of Hydrochloric Acid Concentration on the Leaching Rate of Cerium

Figure 3 shows the effect of hydrochloric acid concentration on the leaching rate of rare earth elements from waste catalysts.

As can be seen from Fig. 3, the leaching rate of cerium is 9.91% when the hydrochloric acid concentration is 1 mol/L. With the increase of hydrochloric acid concentration, the maximum leaching rate of cerium is 44.76% when the hydrochloric acid concentration is 4 mol/L. The high concentration of H^+ reacts with Al_2O_3 in the support that opens the encapsulation of cerium. It promotes the reaction between Cl^- and cerium, improving the leaching rate of cerium.

However, the leaching rate decreases when the concentration of hydrochloric acid continues to increase. Combining Figs. 1 and 2, it can be seen that when HCl concentration reaches 4 mol/L or above, Ce^{3+} concentration decreases below 10%. $CeCl^{2+}$ concentration decreases below 30% and $CeCl_2{}^+$ concentration increases above 50%. Ce^{3+} and $CeCl^{2+}$ can exist stably under this pH condition in the stable zone of aqueous

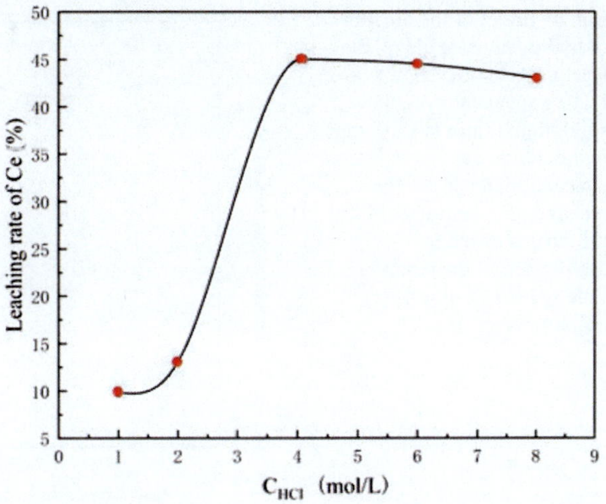

Fig. 3 Effect of hydrochloric acid concentration on the leaching rate of cerium (10.0 g waste catalyst, solid–liquid ratio 1:10, reaction temperature 80 C, stirring speed 400 r/min, reaction time 1 h, hydrogen peroxide consumption 30%, 10 ml). (Color figure online)

solution but $CeCl_2^+$ can not. Therefore,when hydrochloric acid concentration is too high, the leaching rate of rare earth elements will decrease.

Effect of Reaction Temperature on the Leaching Rate of Cerium

Figure 4 investigates the effect of temperature on the leaching rate of rare earth elements from spent catalysts.

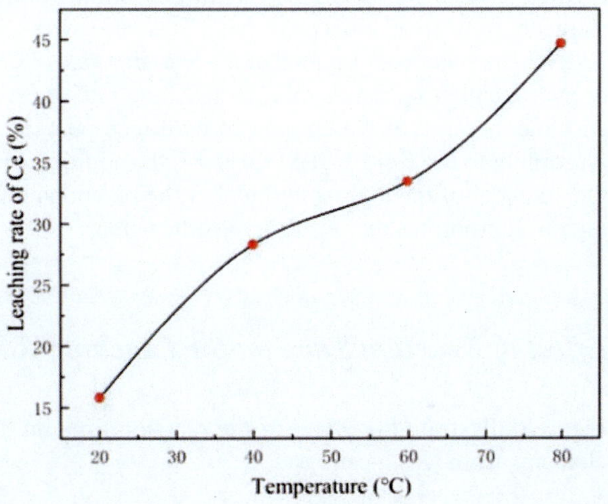

Fig. 4 Effect of reaction temperature on the leaching rate of cerium (10.0 g waste catalyst, solid–liquid ratio 1:10, hydrochloric acid concentration 6 M, stirring speed 400 r/min, reaction time 1 h, hydrogen peroxide consumption 30%, 10 ml). (Color figure online)

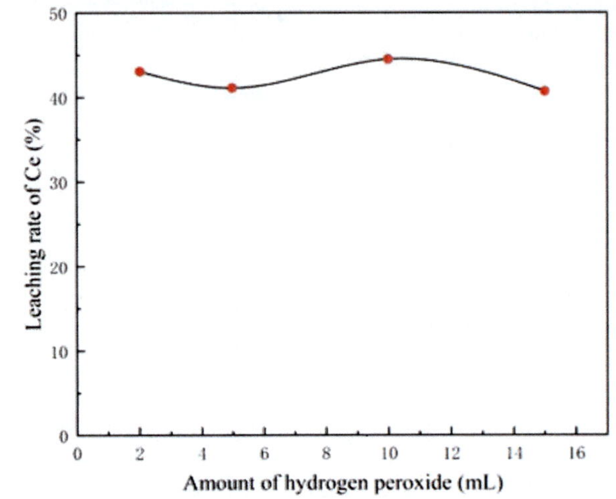

Fig. 5 Effect of the amount of hydrogen peroxide on the leaching rate of cerium (10.0 g spent catalyst, the solid–liquid ratio is 1:10, the hydrochloric acid concentration is 6 M, the reaction temperature is 80 C, the stirring speed is 400r/min, and the reaction time is 1 h.). (Color figure online)

Figure 4 shows that the leaching rate of Ce is 15.77% at 20 °C but 44.42% at 80 °C. Increasing temperature aggravates molecular thermal motion and facilitates the reaction. However, excessive reaction temperature leads to more energy consumption and easy volatilization of hydrochloric acid. Therefore, it is appropriate to leach cerium in hydrochloric acid system with reaction temperature around 80 °C.

Effect of the Amount of Hydrogen Peroxide on the Leaching Rate of Cerium

Figure 5 gives the effect of 30% hydrogen peroxide on the leaching rate of rare earth elements from spent catalysts.

It can be seen from Fig. 5 that the leaching rate of Ce is 43.08% when the amount of hydrogen peroxide is 2 ml. The leaching rate of cerium fluctuates to some extent with the increase of the amount of hydrogen peroxide. But all remain above 40%. According to the Ce–Cl–H_2O diagram, the addition of hydrogen peroxide enhances the potential of the system and makes the reaction conditions fall in the Ce^{3+} stable region. It promotes the reaction to some extent.

Effect of Reaction Time on the Leaching Rate of Cerium

Figure 6 illustrate the effect of the reaction time on the leaching rate of rare earth elements from waste catalysts.

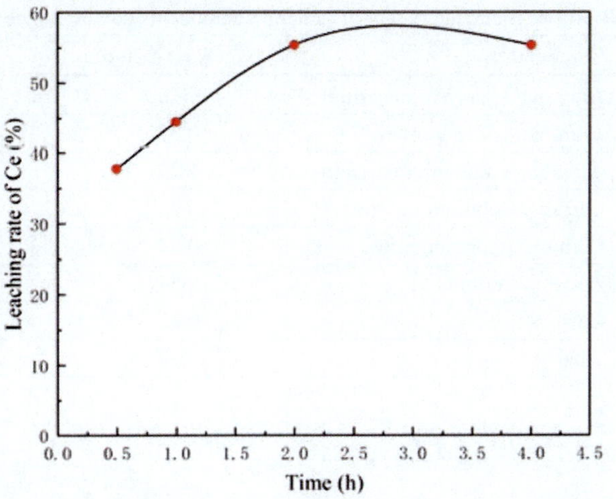

Fig. 6 Effect of reaction time on the leaching rate of cerium (10.0 g spent catalyst, solid–liquid ratio 1:10, hydrochloric acid concentration 6 M, reaction temperature 80 °C, stirring speed 400 r/min, hydrogen peroxide consumption 30%, 10 ml). (Color figure online)

As can be seen from Fig. 6, the leaching rate of cerium is 37.74% when the reaction time is 0.5 h and 55.26% when the reaction time increases to 4 h. The leaching rate of cerium increases with time before 2 h and reaches the maximum of 55.36% after 2 h. As time goes on, the solid–liquid contact becomes more complete and the reaction is more complete. Thus, the leaching rate of cerium is increased. The leaching rate is basically maintained at about 55% after 2 h. Long reaction time will increase energy consumption and production cost. Therefore, 2 h is the best leaching time during the leaching process.

Effect of Liquid–Solid Ratio on the Leaching Rate of Cerium

The effect of liquid–solid ratio on the leaching rate of rare earth elements in the catalyst of waste automobile exhaust purifier was investigated under the conditions of 10.0 g waste catalyst, 6 M hydrochloric acid concentration, 80 °C reaction temperature, 400 r/min stirring speed, 1 h reaction time, and 10 ml of 30% hydrogen peroxide.

The study of this influencing factor only provides experimental data with liquid–solid ratio of 2:1, 3:1, and 10:1. From these three sets of data, it can be found that when the liquid–solid ratio is 2:1, the leaching rate of cerium is 37.93%. While the liquid–solid ratio is 10:1, the leaching rate of cerium is 44.42%. With the increase of liquid–solid ratio, the leaching rate of cerium showed an upward trend.

Table 2 Element content of mineral components containing cerium

Component	Raw materials (%)	Ce (%)	Zr (%)	Ce (g)
Alumina + zirconium cerium alloy 1	15.60	13.23	16.83	0.02063
Alumina + lead cerium	0.37			0.0036
Alumina + zirconium cerium alloy 2	1.29	19.67	6.70	0.00254
Cerium Zirconium mixture	0.42	32.88	29.15	0.00138
Cerium Zirconium oxide	0.37	5.88	61.10	0.00022
Zirconia Cerium	2.54	36.24		0.00921
CeO	0.18	83.72	–	0.00151
Total	20.77			0.0391

Analysis and Discussion

Through the study of hydrochloric acid system, it was found that the leaching effect of cerium in hydrochloric acid was not ideal. Zirconia was added during the preparation of the catalyst so that zirconia could be embedded into the ceria lattice to generate oxygen vacancies, which could improve the ability of oxygen movement. Cerium and zirconium form a solid solution with stable structure. The presence of zirconium increases the high-temperature stability of cerium. Both of them have synergistic effect. Zirconia is stable in nature and slightly soluble in hydrochloric acid, so relevant hypotheses are demonstrated and analyzed.

From the analysis of raw material composition, the percentage of cerium in raw material is 3.91%. It means that 0.0391 g of cerium in 1 g of raw material. After the above experimental demonstration, the leaching rate of cerium in hydrochloric acid system is about 46%. If we calculate it based on the leaching rate of 46%, 1 g raw material can leach 0.018 g Ce in hydrochloric acid. According to the composition analysis of experimental raw materials, the element content of cerium-containing mineral components can be obtained as shown in Table 2.

According to Table 2 and the related literature, it is known that the leaching of cerium is hindered when the content of zirconium is too high. It is assumed that cerium in the components with cerium content higher than zirconium content will be leached. That means alumina + lead cerium, alumina + zirconium cerium alloy 2, cerium-zirconium mixture, zirconia cerium and cerium in CeO can be leached by hydrochloric acid. After calculation, the total amount of cerium that can be dissolved into the solution is 0.018 g, which coincides with the experimental results. It shows that most of the components with low zirconium content may not form zirconium-cerium solid solution while the rest may form zirconium-cerium solid solution. Because of the encapsulation of cerium by zirconium in cerium-zirconium solid solution, it is difficult for cerium oxide to be dissolved into the solution. This ultimately makes the leaching rate of cerium unsatisfactory.

Conclusion

The E-pH diagram of Ce–Cl–H_2O system has been drawn and analysed in this paper. After the experiment of leaching conditions of waste catalyst, the following conclusions are obtained.

(1) Based on the E-pH diagram of Ce–Cl–H_2O system, the introduction of Cl^- has little effect on the stable region of Ce^{4+}. But it has a greater effect on the stable region of Ce^{3+}. The reason may be that it will react with Ce^{3+} to form corresponding complexes, while Ce^{4+} will not react with Cl^-.

(2) Temperature has a greater effect on the E-pH diagram. With the increase of temperature, the stable regions of Ce^{3+} and Ce^{4+} in the system mostly move to the low pH direction. With the increase of pH, ions will react to produce corresponding oxides. Low pH values are required if cerium is to be ionized.

(3) The results of leaching cerium in hydrochloric acid system showed that the maximum leaching rate of cerium was about 46% when the concentration of hydrochloric acid was 4 mol/L, the reaction temperature was 80 C, the amount of hydrogen peroxide was 10 ml, the liquid–solid ratio was 10:1, the reaction time was 2 h, and the stirring speed was 450 r/min.

References

1. Jowitt SM, Werner TT, Weng ZH et al (2018) Recycling of the rare earth elements. Curr Opin Green Sustain Chem 13:1–7
2. Huang XW, Long ZQ, Wang LS et al (2015) Technology development for rare earth cleaner hydrometallurgy in China. Rare Metals 34(4):215–222
3. Chen ZH (2011) Global rare earth resources and scenarios of future rare earth industry. J Rare Earths 29(1):1–6
4. Ferron CJ, Henry P (2016) A review of the recycling of rare earth metals. Can Metall Q 54(4):388–394
5. Christmann P (2014) A forward look into rare earth supply and demand: a role for sedimentary phosphate deposits. Proc Eng 83:19–26
6. Borra CR, Pontikes Y, Binnemans K et al (2015) Leaching of rare earths from bauxite residue (red mud). Miner Eng 76:20–27
7. Yu SQ, Tian HP, Long J (2013) Research progress of low rare earth content fluid catalytic cracking catalyst abroad. Pet Refin Chem Eng 44(8):1–7
8. Ye LS, Lv LL, Yang C, Gao W (2018) Review of rare earth element recovery methods for spent FCC catalyst. Guangzhou Chem Eng 46(10):15–17
9. Zheng WF (2014) Leaching and separation technology of lanthanum and cerium in FCC catalyst. School of Chemistry and Chemical Engineering, Northwest Normal University, Lanzhou
10. Qin YF, Xu T et al (2014) Recovery and comprehensive utilization of rare earth resources in spent catalysts. Rare Earth 35(1):76–81
11. Wu XF, Dong YG, Tong WF et al (2015) Study on the new technology of recovering rare earth elements from spent automobile catalyst. Rare Earths 36(1):101–104

Treatment of Electroless Nickel Plating Wastewater by Ozone Oxidation

Lili Xi, Yahui Sun, Jiann-Yang Hwang, Fujiu Nian, Zhixing Fu, Yang Yang, and Cuiping Huang

Abstract Electroless nickel plating technology is also called chemical nickel plating, which is an important nickel plating method. However, the wastewater produced by electroless nickel plating contains a large amount of phosphorus and nickel. The heavy metal nickel is carcinogenic and has sensitizing effects. It is also toxic to environmental organisms. Phosphorus is a well-known factor that leads to eutrophication of water bodies. How to effectively deal with the waste solution of chemical nickel plating, turning the environmental pollutants into recyclable resources, and reducing the damage to the ecological environment are of great significance. This article applies ozone to the treatment of nickel wastewater. Its success brings the possibility of developing a new method for the treatment of nickel wastewater.

Keywords Ozone oxidation · Water treatment · Nickel wastewater

Introduction

Electroless nickel plating technology has been widely used for its unique advantages. It is a process in which metal ions are reduced to metal elements and plated on the surface of plated parts by reducing agent to provide electrons [1]. In electroless nickel plating bath, hypophosphite is used as reducing agent, nickel sulfate provides nickel ion, and organic acid or salt is used as complexing agent. The main sources of electroless nickel plating wastewater are rinsing water of electroless nickel plating parts and discarded bath solution, and the main components are of high concentration nickel ion, hypophosphite and phosphite, and refractory organic matter [2]. Nickel is a kind of carcinogenic heavy metal substances, but also an expensive metal in

L. Xi · Y. Sun · J.-Y. Hwang (✉) · F. Nian · Z. Fu · Y. Yang · C. Huang
Futianbao Environmental Protection Technology Co., Ltd,, Xian 710000, Shaanxi, China
e-mail: jhwang@mtu.edu

J.-Y. Hwang
Michigan Technological University, Houghton, MI 49931, USA

B. Li et al. (eds.), *Materials Engineering—From Ideas to Practice: An EPD Symposium in Honor of Jiann-Yang Hwang*, The Minerals, Metals & Materials Series,
https://doi.org/10.1007/978-3-030-65241-8_26

shortage. Phosphorus is one of the main pollution factors causing water eutrophication. Therefore, how to effectively treat the phosphorus and nickel in electroless nickel plating wastewater, reduce the environmental pollution and ecological balance damage, has a very important practical significance.

At present, the main treatment methods of nickel plating wastewater are chemical precipitation [3], ion exchange [4], membrane separation [5], and advanced oxidation technology [6]. Although the traditional chemical process is relatively mature and the operation cost is low, the disadvantage is that it is difficult to achieve the effect of advanced treatment, Polymer flocculants are often used, and the efficiency is low and the sludge volume is large. For the complex nickel, because the complexing agent can combine with nickel ion stably, it is difficult to remove it completely by traditional chemical method. Ion exchange method and membrane separation method are difficult to be used on a large scale due to high operation requirements, membrane toxification and saturation regeneration of ion exchanger.

Advanced oxidation technology [7] produces a large number of highly active free radicals (·OH) through various physical and chemical processes, such as light, sound, electricity, and magnetism. The free radicals have a strong oxidation power. Its redox potential is 2.80 v, which is second only to the 2.87 v of fluorine [8]. It is desired to use these highly oxidizing free radicals to oxidize and degrade organic pollutants in water, with strong oxidation and easy control of operating conditions. The common advanced oxidation technologies include Fenton oxidation, electrocatalytic oxidation, and supercritical water oxidation. Liu [9] used Fenton like oxidation to treat electroless nickel plating wastewater, and Xu [10] used electrocatalysis to treat electroless nickel plating wastewater. However, the above processes have some disadvantages, such as large amount of sludge produced by Fenton oxidation, small scale of electrocatalysis technology and high requirements for equipment. Therefore, it is necessary to find a stable and efficient method to treat nickel wastewater.

As an extremely strong oxidant, ozone has the advantages of strong oxidation, simple operation and no secondary pollution, which is widely used in water treatment [11]. The degradation reaction can be divided into direct reaction and indirect reaction. Direct reaction refers to the direct reaction of ozone molecules with pollutants, mainly including redox reaction, cycloaddition reaction, electrophilic substitution reaction, etc., with the characteristics of slow reaction rate and selectivity [12]. Indirect reaction refers to the reaction between hydroxyl radicals (·OH) and pollutants produced by ozone decomposition (or other direct reactions). ·OH ($E^0 = 2.8$ V) has higher redox potential than O_3 ($E^0 = 2.07$ V). Hydroxyl radicals can react with most organic compounds (and some inorganic substances) in water and, therefore, it has unique advantages in the field of water treatment [13].

The treatment of nickel wastewater by ozone catalytic oxidation was studied. The effects of initial pH, ozone dosage and reaction time on the conversion rate of nonpositive phosphorus were investigated. The removal effect of COD (Chemical Oxygen Demand) under the same conditions were also investigated. Chemical precipitation of ozonated wastewater after the optimal reaction conditions was carried out. The results showed that the contents of Δ phosphorus and total nickel were 3.70 mg L^{-1} and 2.05 mg L^{-1}, respectively. After ozone catalytic oxidation treatment, phosphorus and nickel in the system are mainly composed of orthophosphate and free nickel. In the next step, nickel and phosphorus can be recovered by ion exchange and other technologies [14, 15] to realize resource recovery and utilization.

Experimental

Experimental Water Sample

The wastewater used in the experiment is chemical nickel rinsing water from an electroplating factory in Xi'an, Shaanxi Province. The electroplating solution in the workshop is composed of nickel sulfate, sodium hypophosphite, sodium citrate, etc. The organic compounds in the wastewater mainly come from the stabilizers, complexing agents and brighteners added in the plating bath. The main substances include sodium dodecyl sulfonate, citric acid and polyethylene glycol. Nickel ions in the water form complex ions with complexing agents. The results showed that COD, total nickel, total phosphorus, orthophosphate and pH were 309 mg L^{-1}, 161 mg L^{-1}, 147 mg L^{-1}, 0.61 mg L^{-1} and 5.5, respectively.

Ozone Catalytic Oxidation Experiment

The experiment was carried out at room temperature. Ozone was produced by ozone generator and the tail gas was absorbed by potassium iodide solution. The conversion rate of nonpositive phosphorus and the removal rate of COD were analyzed by adjusting the inlet concentration to control the dosage of ozone.

Chemical Precipitation Experiment

Put appropriate amount of ozone treated water into the beaker, place the beaker on the magnetic stirrer, add calcium chloride according to 1.2 times of the theoretical amount, adjust the pH value to 10.5 with NaOH, add PAC (polyaluminum chloride, 3‰) and PAM (polyacrylamide, 1‰) into the beaker, stir at medium speed for 5 min, and then settle down. Filter the supernatant with qualitative filter paper to investigate the changes of total phosphorus and nickel concentration in the water sample.

Analytical Methods and Instruments

COD, total phosphorus, orthophosphate, and other indicators were determined by standard method. The instruments used were visible spectrophotometer (N2, Shanghai Instrument and electric analysis instrument), pH meter (phs-2, Shanghai Leici), nickel was determined by flame atomic absorption spectrophotometry (GB/T 5750.6). The instruments used were atomic absorption spectrophotometer (4510F, Shanghai Instrument and electric analysis instrument), ozone generator (fh-cyj1510c-ys, Shanghai fogha).

Results and Discussion

Effect of Ozone Dosage on Nonpositive Phosphorus and COD

The ozone oxidation experiment was carried out at room temperature. pH was adjusted to 9.0 with NaOH solution, and the reaction time was 90 min. The dosage of ozone was controlled by adjusting the inlet concentration. As shown in Fig. 1, with the increase of ozone dosage, the conversion rate of nonpositive phosphorus in wastewater increases continuously. When the dosage of ozone is 500 mg L^{-1}, the conversion rate of nonpositive phosphorus reaches 85%. If the dosage of ozone continues to increase, the conversion rate will not increase significantly. From Fig. 2, when the ozone dosage is 0–250 mg L^{-1}, the COD removal rate increases rapidly,

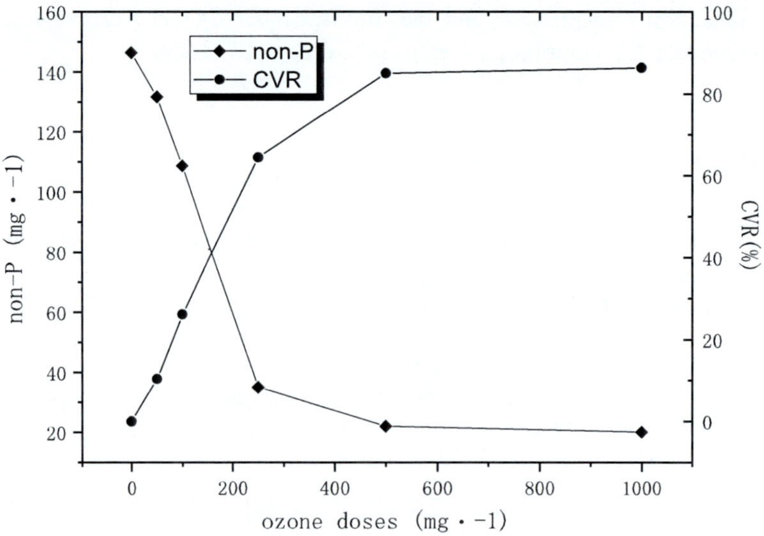

Fig. 1 Effect of ozone dosages on conversion of nonpositive phosphorus

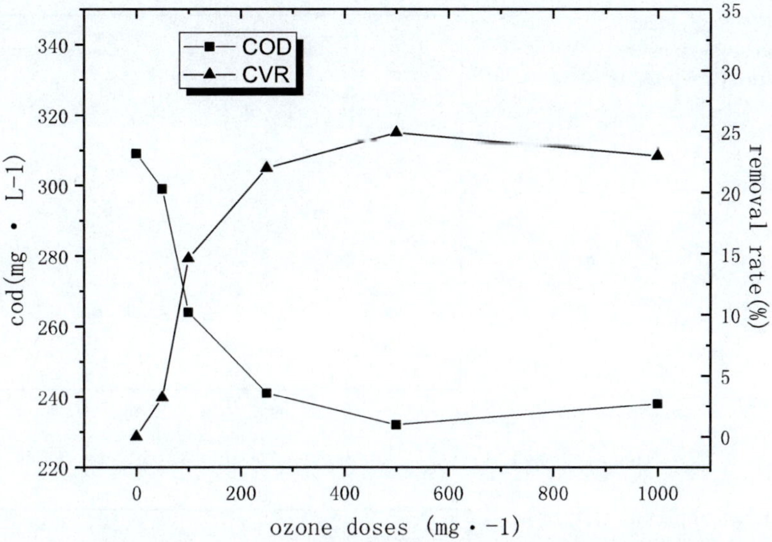

Fig. 2 Effect of ozone dosage on COD removal rate

and with the ozone dosage continues to increase, the removal rate of COD tends to be gentle. The reason is that ozone catalytic oxidation mainly produces ·OH through ozone catalysis, and the reducing substances in water are oxidized by · OH. With the increase of ozone dosage, the content of nonpositive phosphorus and organic matter in water gradually decreases, and reduces the probability of collision with ·OH, which leads to the decline of ozone utilization rate, and the conversion rate of nonpositive phosphorus and the removal rate of COD tend to be flat.

Effect of Initial pH of Solution on Nonpositive Phosphorus and COD

The experiment was carried out under different pH conditions. NaOH and H_2SO_4 were used to adjust the pH of wastewater. The dosage of ozone was 500 mg L^{-1}. The initial pH has a great influence on the ozonation of nonpositive phosphorus. With the increase of pH from 2.0 to 9.0, the conversion rate of nonpositive phosphorus increased from 11.3% to 85%, as shown in Fig. 3. It can be seen from Fig. 4 that when the initial pH is 2, the removal rate of COD is low. When the initial pH is in the range of 5.5–11, the removal rate of COD is close, which is double that of pH 2. This is because ozone is very unstable under alkaline conditions, and it is easy to self decompose. Through a series of spontaneous chain reactions, strong oxidation hydroxyl radicals are generated, mainly indirect oxidation, which is more conducive to the conversion of nonpositive phosphorus. In addition, with the increase of pH of

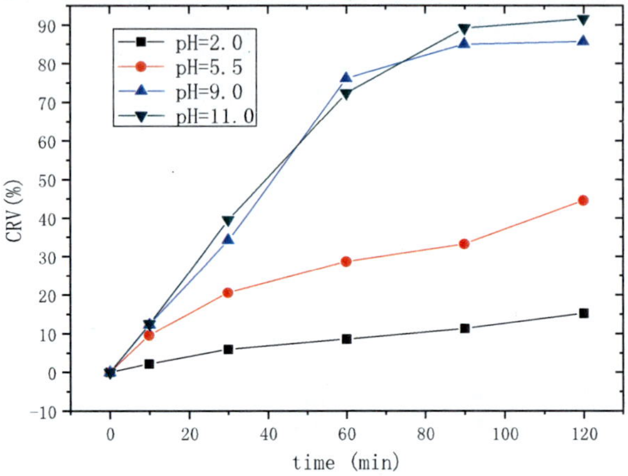

Fig. 3 Effect of pH on conversion of nonpositive phosphorus. (Color figure online)

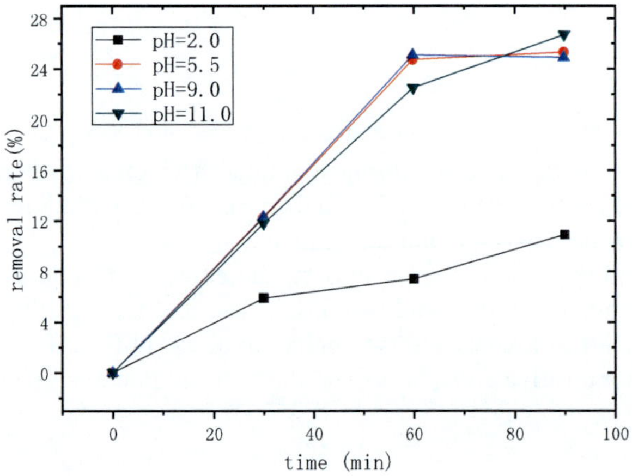

Fig. 4 Effect of pH on COD removal rate. (Color figure online)

wastewater, the concentration of OH^- in water increased, which promoted the ozone decomposition to produce ·OH.

Effect of Reaction Time on Nonpositive Phosphorus and COD

The ozonation experiment was carried out at room temperature. The wastewater pH was 9.0 and the dosage of ozone was 500 mg L^{-1}. The effects of ozonation time on the conversion rate of nonpositive phosphorus and COD removal rate were investigated. The results are shown in Fig. 5.

The conversion rate of nonpositive phosphorus and the removal rate of COD increased with time. The conversion rate of nonpositive phosphorus reached 85% in 90 min and the removal rate of COD reached 25.1% at 60 min. Both of them tended to

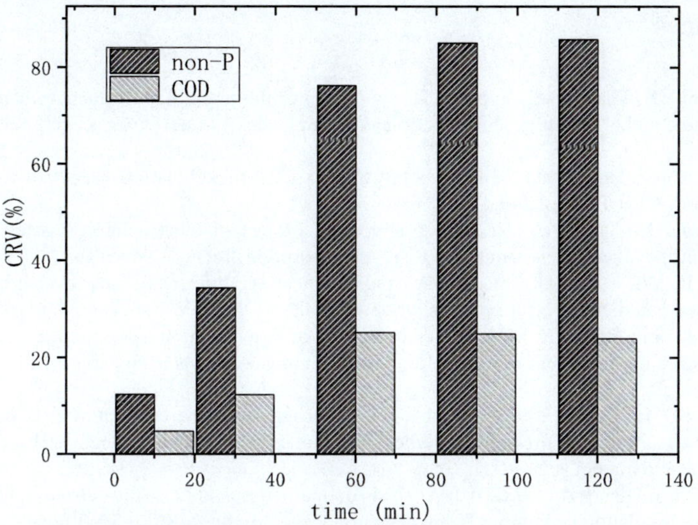

Fig. 5 Effect of reaction time on nonpositive phosphorus and cod

be stable with increasing reaction time. The reason is that the lifetime of ·OH in water is very short, about 10^{-9} s [16]. When the ozone catalytic oxidation time is short, more ·OH is produced per unit time, resulting in a large number of ·OH quenched without participating in the reaction. When the reaction time reaches a certain critical point, the probability of collision between ·OH and reductive substances decreases with the extension of reaction time, which leads to the decrease of ozone utilization.

Conclusions

The optimal reaction conditions were pH of 9, ozone dosage of 500 mg L^{-1} and reaction time of 90 min. The treatment of electroless nickel plating wastewater by ozone catalytic oxidation has obvious effect on oxidation of nonpositive phosphorus and removal of COD. Under the optimal reaction conditions, the conversion rate of nonpositive phosphorus was 85%, the removal rate of COD was 24.9%, and the removal rate of nickel after chemical precipitation was 94.2%. After the wastewater of electroless nickel plating was oxidized by ozone, the solution mainly contains orthophosphate and free nickel ions, which is convenient for the subsequent recovery of nickel and phosphate by ion exchange method. This method can be widely applied to electroless nickel plating bath, recycling resources and turning the environmental pollutants into recyclable resources.

References

1. Bulasara VK, Thakuria H, Uppaluri R et al (2011) Combinatorial performance characteristics of agitated nickel hypophosphite electroless plating baths. J Mater Process Tech 211(9):1488–1499
2. Shao Z, Cai Z, Hu R et al (2014) The study of electroless nickel plating directly on magnesium alloy. Surf Coat Technol 249:42–47
3. Fu L, Dai J, Chen L et al (2018) Engineering project of electroplating wastewater toxic pollutants advanced treatment based on resins adsorption. Technol Water Treat 44(1):128–131
4. Cheng R, Qiu L, Liu G et al (2018) Application of ceramic membrane-reverse osmosis in advanced treatment of electroplating wastewater. China Water Wastewater 34(14):41–45
5. Shih YJ, Lin CP, Huang YH (2013) Application of Fered-Fenton and chemical precipitation process for the treatment of electroless nickel plating wastewater. Separat Purificat Technol 104(Complete):100–105
6. Tu Z, Li D, Li N et al (2002) Present status and progress in the treatment of electroless nickel plating wastewater. In: Proceeding of the 5th annual meeting of China Electrotechnical Society on surface treatment
7. Liu J, Qiu J, Huang A et al (2019) Analysis on mal-degradable organic wastewater treated by advanced oxidation technology. China Environ Protect Indus 248(02):37–39
8. Mantzavinos D, Kassinos D, Parsons SA (2009) Applications of advanced oxidation processes in wastewater treatment. Water Res 43(16):3901
9. Liu Y (2015) Process research in removal of pollutants of electroless nickel plating wastewater. South China University of Technology, Guangzhou
10. Xu G-R, Tian S, Ma Y-Y et al (2017) Treatment of wastewater discharged from electroless nickel plating process by high-voltage pulsed electrocoagulation. Electroplat Finish
11. Huang Y, Jia J (2006) Development of preparation of ozone and its application in water treatment. Hebei Chem Eng Indus
12. Lv J, Zhang D (2011) Application of ozone technology in wastewater treatment. J Water Res Water Eng 22(006):123–126
13. Peng S, Wu D (2019) Research on catalytic ozonation and its application to the advanced treatment of industrial wastewater. Indus Water Treat 039(001):1–7
14. Wu Z, Li J, Wang Y et al (2019) Recovery of nickel from electroless nickel plating wastewater by electrolysis-catalytic reduction method. Electroplat Pollut Control 39(2):55–58
15. Mei X, Cheng H, Zhang Y et al (2013) Selective phosphorus recovery from anaerobic digestion supernatant of sewage sludge by ion exchange. Chin J Environ Eng 7(9):3319–3326
16. Wang H, Jiang Z, Yang H et al (2008) Quantitative analysis of OH generated in electrically assisted photocatalysis. Chin J Environ Eng 2(2):225–228

COD Removal from Electroplating Degreasing Wastewater by UV/H_2O_2 Process

Jianjun Liu, Zhichao Chen, Lili Xi, Jiann-Yang Hwang, and Yong Shi

Abstract Degreasing wastewater is a common type of wastewater produced from surface degreasing process in the electroplating industry. It has the characteristics of high pH, high oil content, and high COD value. This paper studies COD reduction of degreasing wastewater by UV/H_2O_2 oxidation process, and factors that affect COD removal rate. Those include H_2O_2 dosage, feeding method, and pH value. The optimized reaction condition was found to be pH of 3, the molar ratio of H_2O_2 to COD at 1:1 with continuous feeding of H_2O_2. A maximum COD removal rate of 91% can be achieved after 4 h of reaction.

Keywords COD · UV/H_2O_2 oxidation · Degreasing wastewater

Introduction

The electroplating industry as a supporting industry of manufacturing industry is one of the important processing industries in China. In recent years, the manufacturing industry has developed rapidly, and its supporting electroplating industry has also developed rapidly. There are more electroplating enterprises established and different processes utilized. Consequently, electroplating wastewater has complex characteristics, consisting many polluting components such as heavy metals, organics, ammonia, nitrate, cyanide, phosphorous, etc., with large volume and causing serious pollution [1].

Electroplating degreasing solution is one of the stream of wastewater from electroplating industry. It is produced in the pretreatment stage of electroplating process to remove grease, wax, and other impurities from the surface of parts to yield clean parts for a good plating. The solution is mostly alkaline, and the temperature is

J. Liu · Z. Chen · L. Xi · J.-Y. Hwang (✉) · Y. Shi
Futianbao Environmental Protection Technology Co, Ltd, Xian, Shaanxi 710000, China
e-mail: jhwang@mtu.edu

J.-Y. Hwang
Michigan Technological University, Houghton, MI 49931, USA

B. Li et al. (eds.), *Materials Engineering—From Ideas to Practice: An EPD Symposium in Honor of Jiann-Yang Hwang*, The Minerals, Metals & Materials Series,
https://doi.org/10.1007/978-3-030-65241-8_27

between 40 and 90 °C. The solution contains deoiling powder (the main components of which are NaOH, Na_2CO_3), surfactant, corrosion inhibitor, etc., and the oil and wax falling off from the surface of the plating parts. It has the characteristics of high basicity and high organic concentration. The COD (Chemical Oxygen Demand) of deoiling mother liquor can be more than 50,000 mg/L in some cases. After rinsing the residual mother liquor off from the surface of the plating parts, secondary wastewater is produced with diluted concentrations of pollutants.

At present, activated carbon adsorption, biological treatment, and advanced oxidation have been studied to remove organic matters, the COD [2]. Activated carbon adsorption can remove most of the organic matter, but the adsorption capacity of activated carbon is small, the cost is high, and the adsorption materials need to be regenerated repeatedly, and the regeneration process is complex and difficult to operate [2]. Biological treatment is rarely used in electroplating wastewater, since its high salt content will inhibit microbial growth. Advanced oxidation has the characteristics of strong oxidation and wide applications. It can remove most organic matter in water and reduce COD content. The common oxidants include O_3, H_2O_2, Fenton reagent, etc. The oxidation efficiency of O_3 is low and easy to cause air pollution [3]. Fenton process has large dosage of chemicals and produces a large amount of sludge, which needs further treatment. H_2O_2 does not produce sludge and has high oxidation efficiency. In order to improve COD removal rate, H_2O_2 can also be combined with UV light, catalyst, and O_3. In this study, the electroplating degreasing wastewater from an enterprise was collected. The combination of H_2O_2 and ultraviolet light was selected to investigate the optimal process conditions. Because of the complex composition of degreasing wastewater, CODcr。was used as the index in the whole experiment.

Experimental Device and Method

Water Quality

Degreasing waste liquid is taken from the alkaline degreasing mother liquor of a zinc plating enterprise. The waste liquid is turbid, light gray, and a small amount of white precipitate is produced after standing. Its main components include NaOH, Na_2CO_3, anionic surfactant, corrosion inhibitor, oil, and a small amount of zwitterionic metal ions (Zn, Al). The main water quality parameters monitored in this experiment are shown in Table 1.

Table 1 Water quality parameters of deoiling waste liquid

Water quality parameters	Determination results
pH	13.65
CODcr,mg/L	5934
Oil, mg/L	666

Experimental Methods

The experimental reagents were 30% H_2O_2, NaOH, MnO_2, and H_2SO_4. The schematics of the experimental setup is shown in Fig. 1. The UV lamp power is 23 W, the wavelength is 254 nm, the length is 420 mm, and the diameter is 30 mm. The experimental vessel is a 1.0 L ordinary glass measuring cylinder, 390 mm in length and 80 mm in inner diameter. The aeration device is a laboratory small microporous aerator, and the lampshade is made of tin paper, which is wrapped on the surface of the measuring cylinder.

During the experiment, 1000 mL of deoiling wastewater was added into the container, the pH of the solution was adjusted with NaOH and H_2SO_4, and then 30% H_2O_2 was added. The UV lamp and aeration device were turned on, and samples were collected for inspection every 30 min.

COD of samples was determined by potassium dichromate method. After UV/H_2O_2 treatment, the residual H_2O_2 in water will interfere with COD_{Cr}, so it is necessary to pretreat the solution before determination. The specific steps are as follows: 100 mL solution was obtained and pH was adjusted to 8–9. The solution was added 1 g MnO_2 and stirred for 1 h, followed by centrifugation of 2000 r/min for 10 min. The controlled experiment was conducted without MnO_2 and centrifugation process.

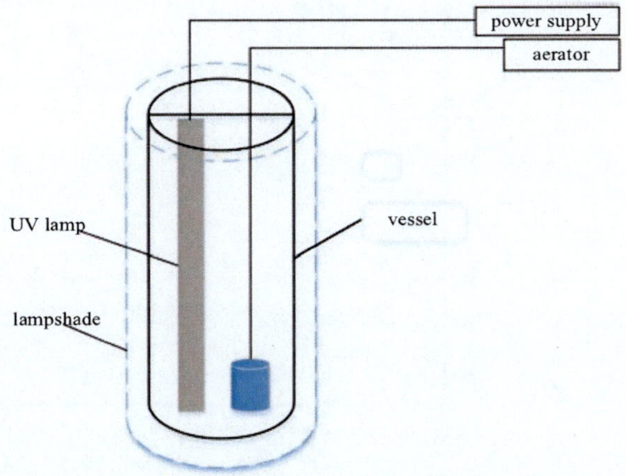

Fig. 1 Schematic diagram of experimental device. (Color figure online)

Results and Discussion

Effect of PH Value on COD Degradation

In general, the redox potential of the same redox agent in different pH solutions will be very different. At the same time, H_2O_2 is more sensitive to the pH of the solution, so the change of pH will have a great impact on the ability of UV/H_2O_2 process to oxidize the organic matters.

In the experiment, the pH of the original solution was controlled at 3.0, 6.0, 9.0, and 13.0 (the initial pH of the original solution was about 13.0). The dosage of H_2O_2 was added at one time according to the molar ratio of COD: H_2O_2 for 5 h. The experimental results are shown in Fig. 2.

It can be seen from Fig. 2 that when the initial pH value increases from 3 to 13, the COD value of wastewater decreases, and the COD removal rate (Crr) reaches the maximum when the pH value is 3. The results show that the optimal pH value of the reaction system is 3, and the maximum COD removal rate is 84%. The reason is that H_2O_2 itself is a weak acid, which is easy to ionize under alkaline conditions, while the consumption of H_2O_2 is more stable under neutral and acidic conditions [4]. In addition, when UV/H_2O_2 oxidation process is applied to the actual water treatment, it is necessary to consider the quenching of OH hydroxyl radical by other components in the water, including Br^-, $CO_3{}^{2-}$, and $HCO_3{}^-$ [4].

On the other hand, when the reaction time continued to extend, the COD removal rate did not increase significantly. This was due to the gradual consumption of H_2O_2 in the reaction process. After reaching a certain time, the concentration of H_2O_2 in

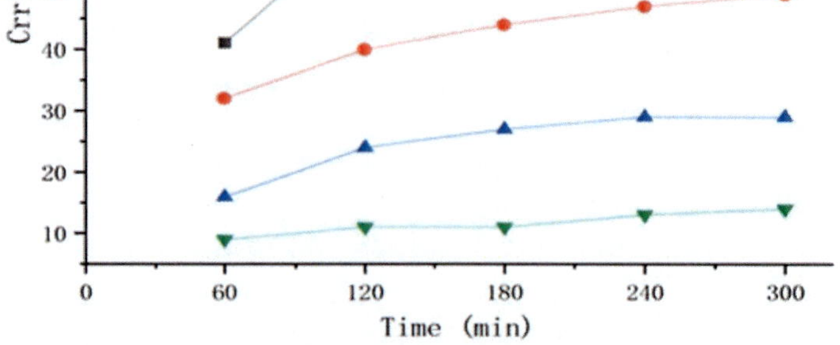

Fig. 2 Change of COD removal rate under different pH conditions. (Color figure online)

solution was too low, and the COD removal rate did not increase significantly when the reaction time continued to extend.

Effect of H_2O_2 Dosage and Mode on COD Degradation

H_2O_2 has an important effect on the initial concentration of UV/H_2O_2. Generally speaking, the increase of initial concentration of H_2O_2 will increase the concentration of OH in the solution, thus increasing the reaction rate and target removal rate [5]. However, there is no linear relationship between the initial concentration of H_2O_2 and the target removal rate, and there is an optimal range for the dosage of H_2O_2.

In this experiment, the optimum pH and reaction time determined earlier were used. The pH of the stock solution was controlled to be 3.0. The molar ratio of H_2O_2 to COD was 0.5, 1, and 2, respectively. One group was to add H_2O_2, at one time, and the other half was added with half the amount of H_2O_2. After reaction for 1 h, the other half was added continuously by peristaltic pump within 3 h and irradiated with 23 W UV lamp for 240 min. The COD value is determined by sampling every 60 min. The experimental results are shown in Fig. 3.

The experimental results show that the COD removal rate does not increase continuously with the increase of H_2O_2 dosage, but the highest COD removal rate is obtained when the molar ratio is 1:1. This is consistent with gkenf et al. [6] that there is a critical H_2O_2 dosage in UV/H_2O_2 system. There are two reasons: (1) during the reaction process, H_2O_2 is consumed continuously and decomposes

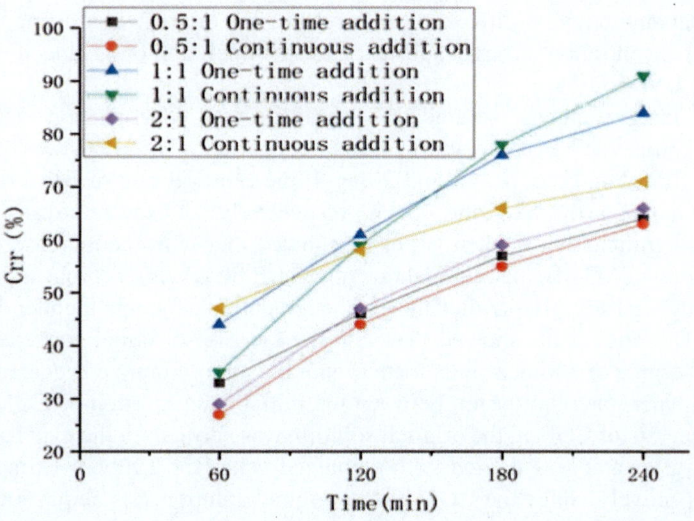

Fig. 3 Change of COD removal rate under different H_2O_2 addition amount and adding mode. (Color figure online)

itself, so that enough H_2O_2 can produce enough OH [6] to promote the reaction to continue; (2) excessive H_2O_2 acts as a free radical aggressive agent to capture OH generation by-products in the solution, thus inhibiting the degradation efficiency of target compounds. H_2O_2 is a strong free radical inhibitor. The reaction rate constant of H_2O_2 with OH is 2.7×10^7L/(mol·s) [7]. The related side reaction process is expressed by formula (1) and (2) [7]

$$H_2O_2 + OH \rightarrow \cdot HO_2^- + H_2O \tag{1}$$

$$2HO_2^- \rightarrow H_2O_2 + O_2 \tag{2}$$

Due to the long reaction time, if H_2O_2 is added at one time, a large amount of H_2O_2 may decompose and inhibit the production of OH in the initial stage. It is difficult to have enough H_2O_2 in the system. Therefore, we try to add H_2O_2 continuously. The results showed that when the molar ratio of H_2O_2: COD was 1:1 and H_2O_2 was added continuously, the removal rate of COD could reach 91%.

Effect of Initial Target Concentration

For the ordinary reaction system, increasing the initial concentration of reactants can increase the reaction rate. However, for the system involving optical reaction, the influence of the initial concentration of target is more complex. In the actual production and application, the concentration of organic matter in raw water also has regional and temporal differences. Therefore, it is of great significance to study the initial concentration of target substance for the operation of advanced oxidation process of UV/H_2O_2.

Xinxin Feng et al. [8] studied the degradation of benzophenone (BP-3) by UV/H_2O_2 process. When the initial mass concentration of BP-3 was gradually increased from 5mggl to 10, 15 and 20mggl, the removal rate of BP-3 decreased from 94.7% to 74.0%, 62.9 and 45.4%, respectively, and the removal efficiency decreased significantly. Stephen [9] et al. showed that with the increase of initial concentration of MTBE, the removal rate of MTBE by UV/H_2O_2 process decreased significantly. Lu Liqin [10] studied the photodegradation of orange II under the action of UV/H_2O_2. The results showed that with the increase of initial concentration of H_2O_2, the degree of photochromism and mineralization of orange II decreased.

In this paper, the relationship between the initial concentration of COD and the degradation rate of COD in the original solution was verified. In the experiment, the deoiling wastewater was diluted with deionized water for 2 times, 3 times, and 6 times, respectively. Since the COD of the original solution was about 6000 mg/L, several groups of diluted solutions were, respectively, recorded as n = 3000, n = 2000, and n = 1000. The initial pH of each group of solutions was controlled to be 3, and H_2O_2 was added at one time according to the molar ratio of H_2O_2: COD =

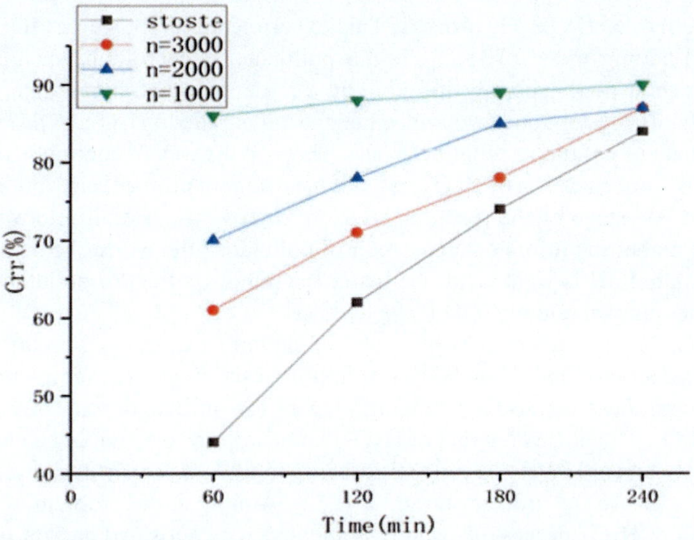

Fig. 4 Change of COD removal rate under different initial concentrations. (Color figure online)

1:1, and irradiated with ultraviolet lamp for 4 h, and the samples were detected every 60 min. The experimental results are shown in Fig. 4.

It can be seen from Fig. 4 that the experimental results are in good agreement with those of Xinxin Feng and others. With the increase of the initial COD concentration in the solution, the COD removal rate gradually decreases, and this phenomenon is more obvious at the initial stage of the reaction. This is different from the general chemical reaction. The reason is that with the increase of the initial pollutant concentration, the transmittance of the solution will decrease, that is, the ability of UV lamp to penetrate the solution decreases. The effective UV light received by H_2O_2 decreases, thus reducing the formation of OH. The treatment capacity of the whole process will decrease, and the degradation efficiency of pollutants will decrease accordingly [11]. However, in the actual process, reducing the initial concentration of target means that the treatment capacity of wastewater will increase and the energy consumption will increase. Therefore, the actual operability of reducing the initial concentration of target to increase the COD removal rate is still insufficient.

Summary and Prospect

UV/H_2O_2 mainly depends on the strong oxidation of OH to degrade organic matter. The treatment process is affected by many factors, such as UV dose, H_2O_2 concentration, initial pollutant concentration, initial pH, and other factors that will affect the treatment effect of UV/H_2O_2 process. This experiment mainly studies the concentration and adding mode of H_2O_2, initial pollutant concentration and initial pH. The results show that, generally speaking, in a certain concentration range, with the increase of initial H_2O_2 concentration, when the molar ratio of H_2O_2: COD = 1:1, the removal effect of pollutants will be the best. Through the way of continuous addition of H_2O_2, the concentration of H_2O_2 can still have a more uniform distribution in the middle and late stage of the reaction, so as to improve the overall removal rate of COD. The higher the initial concentration of pollutants, the worse the degradation. When the initial pH is weak acid, the better the removal effect of pollutants; when pH = 3, the removal rate of COD is the highest.

UV/H_2O_2 has the advantages of strong oxidation capacity, high treatment efficiency, no selectivity, and no secondary pollution. It shows great development potential. However, there are some problems such as low utilization rate of H_2O_2 and residual H_2O_2. The utilization rate of H_2O_2 can be improved by adding catalyst. It is the focus of current UV/H_2O_2 process research to find an economical and efficient catalyst to improve the utilization rate of H_2O_2. Aiming at the problem of residual H_2O_2, the UV/H_2O_2 process is usually combined with activated carbon (GAC) to remove residual H_2O_2 through activated carbon reaction.

References

1. Liu J, Yan L, Bai W et al (2011) Research progress of advanced oxidation technology in water treatment. Water Treat Technol 37(3):11–17
2. Li H, Shen Y, Li H et al (2011) Present situation and research progress of advanced oxidation technology of wastewater. Water Treat Technol 37(6):6–9
3. Li H, Lu X, Mei P (2006) Research progress of advanced oxidation technology for oilfield wastewater treatment. China Resour Compr Util 2:42–46
4. Liu Y, Zhang J (2011) Recent research progress on reaction mechanism and influencing factors of UV/H_2O_2 advanced oxidation process. Energy Chem Eng 32(3):18–24
5. Wang H, Zhong Y, Shi L et al (2014) Photoconversion of phenol in ice under UV/H_2O_2 and its relationship with hydroxyl radical. J Jilin University: Science Edition 52(1):144–150
6. Gkenf Z (2006) Pre-ozonation of aqueous azo dye (Acid Red -151) followed by activated sludge process. Chem Eng Journal 123(3):109–111
7. Tao S, Hu J, Wang J et al (2020) Study on treatment of organic pollutants in gas field water by advanced oxidation method. Pet Nat Gas Chem Ind 3:129–134
8. Feng X, Durden, Guo Y et al (2015) Kinetics and influencing factors of benzophenone degradation by UV/H_2O_2. Environ Sci (6):2129–2137
9. Cater SR, Stefan MI, Bolton JR et al (2000) UV/H_2O_2 Treatment of Methyl tert-butyl ether in contaminated waters. Environ Sci Technol 34(4):659–662
10. Lu L, Wang Y (2012) Photodegradation of orange II under UV/H_2O_2. Chem Bioeng 29(1):79–80
11. Jin H, Yue J, Wang H et al (2016) Research progress on removal of organic pollutants in water by UV/H_2O_2 advanced oxidation process. Urban Water Supply 12(2):12–15

Part VI
Poster Session

Effect of Antioxidant on Resistance to Ammonia Erosion of Carbon Sleeve in Continuous Annealing Furnace for Low-Temperature Grain-Oriented Silicon Steel Production

Mingsheng He, Jing Zhang, and Yong Lei

Abstract In continuous annealing furnace for low-temperature grain-oriented silicon steel production, carbon sleeve is used as one kind of the best hearth rolls to support and convey steel strip. However, the surface of carbon sleeve in nitriding zone is seriously corroded after a period of time, the surface roughness of carbon sleeve increases, the edge wears and even buildups appear, which seriously affects the surface quality of products. Based on the working conditions of carbon sleeve, the causes and mechanism of corrosion on carbon sleeve by ammonia, and effects of antioxidants on surface quality of low-temperature grain-oriented silicon steel are discussed. In terms of the resistance to ammonia, phosphate is not a good antioxidant for carbon sleeve in continuous annealing furnace for low temperature grain-oriented silicon steel production.

Keywords Carbon sleeve · Resistance to ammonia · Low temperature grain-oriented silicon steel · Continuous annealing · Nitriding

Introduction

Due to the low heating temperature and low production cost, the low-temperature grain-oriented silicon steel has been paid more and more attention. After low-temperature heating, hot rolling, and cold rolling, the steel strip must be treated by continuous decarbonization annealing and nitriding in an atmosphere with NH_3. Graphite, which has the properties of low hardness, good self-lubricity, small thermal expansion coefficient, and excellent mechanical strength at high temperature, is used as one kind of the best hearth rolls to support and convey silicon steel strip in continuous annealing furnace. Impregnating phosphates is an economical, practical, and

M. He (✉)
R&D Center of Wuhan Iron & Steel Co., Ltd., Wuhan 430080, China
e-mail: hms03@tsinghua.org.cn

J. Zhang · Y. Lei
Silicon Steel Division of Wuhan Iron & Steel Co., Ltd., Wuhan 430080, China

B. Li et al. (eds.), *Materials Engineering—From Ideas to Practice: An EPD Symposium in Honor of Jiann-Yang Hwang*, The Minerals, Metals & Materials Series,
https://doi.org/10.1007/978-3-030-65241-8_28

effective method to improve the oxidation resistance of graphite materials [1, 2]. The antioxidant of carbon sheath is generally acid phosphate, and the most commonly used is aluminum dihydrogen phosphate [3]. However, the surface of carbon sleeve in nitriding zone is seriously corroded after a period of time, the surface roughness of carbon sleeve increases, the edge wears and even buildups appear, which seriously affects production efficiency and surface quality of the products. The normal maintenance cycle is six months. However, due to the erosion and wear of carbon sleeve, sometimes downtimes occur approximately every two to three months for maintenance and replacement of the carbon sleeve. It is unknown that the real reason of carbon sleeve is eroded by ammonia in continuous annealing furnace for low-temperature grain-oriented silicon steel production. In this paper, the effect of antioxidant on resistance to ammonia erosion of carbon sleeve is discussed.

Working Conditions of Carbon Sleeve

Temperature

In nitriding zone of the continuous annealing furnace, the temperature was generally in the range of 600–950 °C. Graphite can react with oxygen, water vapor in this temperature range. Therefore, oxidation of the carbon sleeve is inevitable during the production of continuous annealing for low-temperature grain-oriented silicon steel production.

Atmosphere

In nitriding zone of the continuous annealing furnace, it is weak oxidizing atmosphere with H_2–N_2–H_2O, or weak basic and weak oxidation atmosphere with H_2–N_2–H_2O–NH_3. In general, the concentration of H_2 is 40–70%, the concentration of N_2 is 30–60%, and the concentration of NH_3 is different from enterprises, lines and products at home and abroad.

Dew Point

In nitriding zone, the dew point is generally controlled in the range of 30–80 °C. After the steel strip enters the nitriding zone, the decarburization process has been completed. The structure of oxide layer on the surface of steel strip is controlled by the dew point in the furnace.

Running Speed

In the normal production process, the speed of the carbon sleeve is 50–120 m/min. Of course, the operating speeds of different enterprises, different lines, and different grain-oriented silicon steel products at home and abroad are different.

Carbon Sleeves After Use

As can be seen from Fig. 1, the main damage forms of carbon sleeve are erosion, oxidation, and wear. After using the carbon sleeve online for about 3 months, there are many pits and holes on the surface of carbon sleeve in nitriding section, some of which are peeled off and some are seriously worn. Especially after the carbon sleeve is worn to dumbbell shape, it will seriously affect the strip shape and surface quality of the product.

Causes Erosion of Carbon Sleeve

At normal temperature and pressure, graphite does not make any chemical reactions except the long-term immersion in nitric acid, hydrofluoric acid or in fluorine, bromine and other strong oxidizing atmosphere leads to form intercalation compounds slowly. However, due to complex procedures and processes of producing carbon sleeves together with porous material itself. Acid phosphate is the most commonly used antioxidant of impregnation treatment for carbon sleeve, but acid phosphates are easy to absorb moisture and deliquesce, which results in reducing or even losing the resistance to oxidation of the carbon sleeve. As a result, the resistance to oxidation of carbon sleeve becomes worse and the service life becomes shorter.

Aluminum dihydrogen phosphate is the most commonly used antioxidant, while ammonia is an alkaline gas. At the same time, the dew point in the continuous annealing furnace is high. Therefore, aluminum dihydrogen phosphate can react with ammonia or ammonia water as follows:

$$2Al(H_2PO_4)_3 + 6NH_3 = 3(NH_4)_2HPO_4 + Al_2(HPO_4)_3$$

$$Al(H_2PO_4)_3 + 6NH_3 = 2(NH_4)_3PO_4 + AlPO_4$$

$$2Al(H_2PO_4)_3 + 6NH_3 \cdot H_2O = 3(NH_4)_2HPO_4 + Al_2(HPO_4)_3 + 6H_2O$$

$$Al(H_2PO_4)_3 + 6NH_3 \cdot H_2O = 2(NH_4)_3PO_4 + AlPO_4 + 6H_2O$$

Generally speaking, graphitization degree of carbon sleeve is over 70%. Although the surface of carbon sleeve is smooth after soaking of various chemical substances and processing, graphite begins to react with water vapor over 700 °C. Therefore,

Fig. 1 Photos of carbon sleeves after use. (Color figure online)

the following reaction between C and H_2O is likely to occur above 700 °C:

$$C(s) + H_2O(g) = CO(g) + H_2(g)$$
$$CO(g) + H_2O(g) = CO_2(g) + H_2(g)$$

Phosphate itself is an inorganic high-temperature binder. Because of ammonia, high temperature, and high dew point in nitriding zone of the continuous annealing furnace, the antioxidants on the surface of carbon sleeve are continuously consumed, which leads to the loss of oxidation resistance and bonding function, and results in local graphite particles falling off or being oxidized and forming erosion pits on the surface of carbon sleeve. The results show that the surface roughness of carbon sleeve increases, and the continuous oxidation of water vapor further accelerates the oxidation wear of carbon sleeve. Finally, the service life of carbon sleeve is significantly shortened, which seriously affects the production efficiency and product quality.

Conclusions

There are two main reasons for the erosion and damage of carbon sleeve in nitriding zone of the continuous annealing furnace for low-temperature grain-oriented silicon steel production: one is the reaction of phosphate antioxidants with ammonia; the other is that the high dew point in annealing furnace leads to oxidation wear of carbon sleeve. In terms of the resistance to ammonia, phosphate is not a good antioxidant for carbon sleeve. It is necessary to develop low phosphate or non-phosphate antioxidants.

References

1. Maier CR, Jones LE (2005) (2005) The influence of aluminum phosphates on graphite oxidation. Carbon 43:2272–2276
2. He MS, Wang XK, Gong XC et al (2019) Effect of phosphate antioxidant on resisting to buildups formation of carbon sleeves in continuous annealing furnace for silicon steel production. Charact Miner Metals Mater 2019:605–615
3. He MS, Peng SJ, Xue GF et al (2015) Cause analysis on buildup formation of carbon sleeve in continuous annealing furnace for non-oriented silicon steel produced by CSP process. Charact Miner Metals Mater 2015:587–593

Evaluation of Ballistic Behavior by Residual Velocity of Epoxy Composite Reinforced with Sisal Fabric After UV Radiation Exposure

Michelle Souza Oliveira, Fernanda Santos da Luz, Lucio Nascimento, and Sergio Neves Monteiro

Abstract The objective of this work was to evaluate the influence of photodegradation by ultraviolet radiation in composites of epoxy matrix reinforced with sisal fabric in the energy absorption capacity against shots of 0.22-gauge lead projectile. The ballistic tests were performed at subsonic speed using a 150-bar compressed air draft system. The ballistic efficiency was evaluated using the residual velocity technique. Fourier transform infrared spectroscopy (FTIR) was also performed to observe changes after radiation exposure. Exposure to UV radiation, both at 75–225 h, caused changes in the color of the composite plates, in addition to optimizing the energy absorption capacity in level I events by NIJ 0101.04, absorbing about 93% of the projectile energy, showing to be a very easy and fast technique for improving the ballistic properties of composites which in turn are more economically viable than the commonly used synthetic materials.

Keywords Ballistic performance · UV radiation · Aging materials · FTIR · Natural fabrics · Polymer composites

Introduction

Degradation of polymers is any destructive reaction that can be caused by chemical, physical, or mechanical agents or simultaneously. Degradation causes an irreversible change in the properties of polymeric materials, being evidenced by the progressive deterioration of these properties, including the visual aspect [1–4]. In polymeric composites, degradation of the polymeric matrix, reinforcement, or the system as a whole can occur. In the absence of light and at room temperature, most polymers are stable for long periods of time. The degradation occurs due to the phenomena of photolysis, chemical decomposition caused by light, and photo-oxidation. The degradation in natural fibers occurs initially in lignin, which is responsible for the color change of the fiber. In comparison to lignin, cellulose is much less susceptible

M. S. Oliveira (✉) · F. S. da Luz · L. Nascimento · S. N. Monteiro
Military Institute of Engineering – IME, Rio de Janeiro, Brazil
e-mail: oliveirasmichelle@gmail.com

B. Li et al. (eds.), *Materials Engineering—From Ideas to Practice: An EPD Symposium in Honor of Jiann-Yang Hwang*, The Minerals, Metals & Materials Series,
https://doi.org/10.1007/978-3-030-65241-8_29

to UV degradation, while hemicellulose and cellulose are more susceptible to high temperatures [5].

Sisal is a vegetable fiber from a coastal city in Yucatan, Mexico, belonging to the genus *Agave sisalana*. In Brazil, this fiber is grown in semi-arid regions, with the state of Bahia contributing 85% of total world production. It is used in handicrafts, forage baling, production of tequila, and others. Sisal, known to being a source of income and employment for a large contingent of workers, is an important agent for fixing man to the Northeastern semi-arid region, given that, in some of these regions, the only cultivation alternative with satisfactory economic results [6–9]. Sisal is a rigid, durable vegetable fiber that has one of the highest tensile strengths compared to other natural fibers [7].

In order to use composites reinforced with natural fibers in ballistic armors, it's useful to evaluate the lifetime of them. One way to analyze this is to study its durability when exposed to severe environmental conditions. In this scope, the present work aims to assess the physical degradation of these composites when exposed to ultraviolet radiation. To simulate this ultraviolet degradation, the surfaces of these composites were exposed to UV rays in certain amounts of time. Usually, according to the literature, degradation causes a yellowing of the polymer surface [10, 11]. In the case of natural lignocellulosic fibers, this modification is attributed to chemical groups belonging to lignin that undergo electronic transitions when absorbing light in the ultraviolet range (below 350 nm), promoting the degradation and premature aging of the fiber, as degradation and reduction occur of the molar mass. The ballistic evaluation is done with the objective of measuring the total energy absorbed by the layer composed of natural fibers, by evaluating the residual velocity after the impact of the projectile.

Materials and Methods

The sisal fabric was purchased from the company Sisalsul. The material has 100% sisal composition, arranged in rolls, and is characterized by having a simple weave, as shown in Fig. 1. The fabrics were cut to 12 cm × 15 cm. In sequence, fabrics were placed on a stove for 24 h to remove the moisture. For the production of composites, the proportion of 30% by volume of sisal tissue was considered. Since the density of sisal varies from 1.26 to 1.5 g/cm^3, the value used was the average of the densities presented (1.38 g/cm^3).

The 120 × 150 mm layers of fabric were dried at 60 °C in an air-oven for 24 h, in order to reduce the inherent moisture. The epoxy resin used was the commercial epoxy type diglycidyl ether of bisphenol A (DGEBA) hardener with triethylenetetramine (TETA), both distributed by Epoxyfiber, Brazil. The DGEBA/TETA epoxy with stoichiometric phr 13 amounts of hardener.

For the preparation of composites, the proportion used for it was 30 vol% of sisal fabric. A pressure of about 3 MPa was applied to the metallic mold. The laminate

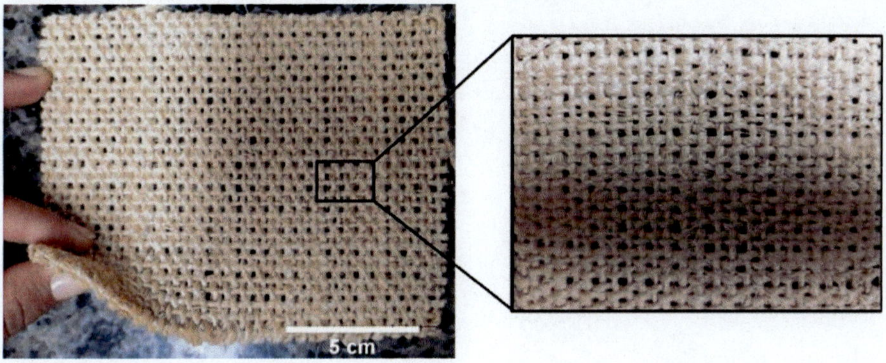

Fig. 1 Sisal fabric in simple weft configuration. (Color figure online)

fabric composite plate, with 3 mm of thickness, was cured at room temperature for 24 h still under pressure.

For ballistic tests, the Gunpower SSS compressed air rifle was used with a noise suppressor, the Standard Weapons. The projectile was for 0.22-gauge lead with an estimated mass of 3.3 g. The impact speed was measured using an Air Chrony MK3 ballistic chronograph with an accuracy of 0.15 m/s.

In ballistic tests, the air rifle was positioned 5 m away from the target and aligned perpendicularly to the rifle. Suppressor noise was used to increase the stability of the projectile at the outlet of the air rifle, reducing the whirlwind caused by the exhaust. The ballistic chronograph was placed 10 cm from the noise suppressor outlet. Figure 2 shows the representative diagram of the ballistic test assembly.

The samples were separated into four evaluation groups, according to Table 1. In each of the composites, five shots were deferred.

The composites were exposed to ultraviolet radiation according to the ASTM G154 standard [12]. The test was carried out at the Environmental Testing Laboratory for Polymers—IME/RJ, in a chamber of an accelerated aging system for non-metals—Ultraviolet B model C-UV from Comexim. The fluorescent lamps used as a source of UV radiation have an intensity of 40 W, operating at 280/320 nm, with a peak at 313 nm. Fourier transform infrared spectroscopy (FTIR) was performed using a Thermo Scientific Smart ITR 50 FT-IR spectrometer, Nicolet model, in the region between 400 and 4000 cm^{-1}. Transmission spectra were obtained with a resolution of 4 cm^{-1} and 64 scans in each test, to obtain a good signal/noise ratio.

Results and Discussion

The results of the composite shots can be seen in Table 2.

The main form of fracture observed was by delamination of the composite, a loss of 3% of the total mass of the composite was recorded after the shots in the NE;

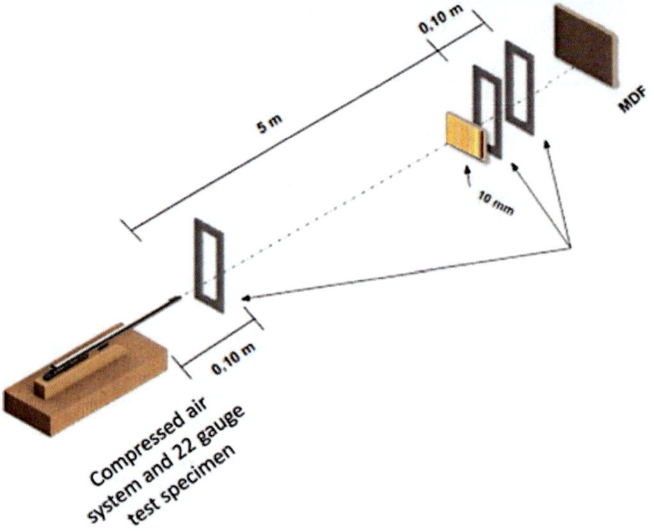

Fig. 2 Schematically ballistic test with compressed air system, 22-gauge test piece, noise suppressor, ballistic chronograph, MDF bulkhead, distance of 5 m between the suppressor end and the bulkhead, and a distance of 10 cm between the end of the suppressor and the chronograph. (Color figure online)

Table 1 Evaluation groups and exposure conditions

Specimens group	Conditions
NE	Non-exposed
UV-75 h	Exposed to 75 h of ultraviolet radiation-B
UV-160 h	Exposed to 160 h of ultraviolet radiation-B
UV-225 h	Exposed to 225 h of ultraviolet radiation-B

Table 2 Energy absorbed from composites after exposure to UV radiation

	Group Control (GC)	75 h	160 h	225 h
Average	91.78	114.53	–	90.89
Standard deviation	24.14	7.13	–	41.04

similarly, a loss of mass of 5% was recorded for the UV-75 h composite, and the other groups had an increase in mass, as both captured the projectile.

For the NE, without exposure to UV radiation, the absorption of 8% of the initial energy of the projectile was obtained, the UV-75 h composite obtained the absorption of 2% of the initial energy of the projectile, with only three valid shots of a total of three shots. One can notice that, for UV-160 h composites (Fig. 3), did not register residual velocity, absorbing all the energy of the projectile. There was no complete

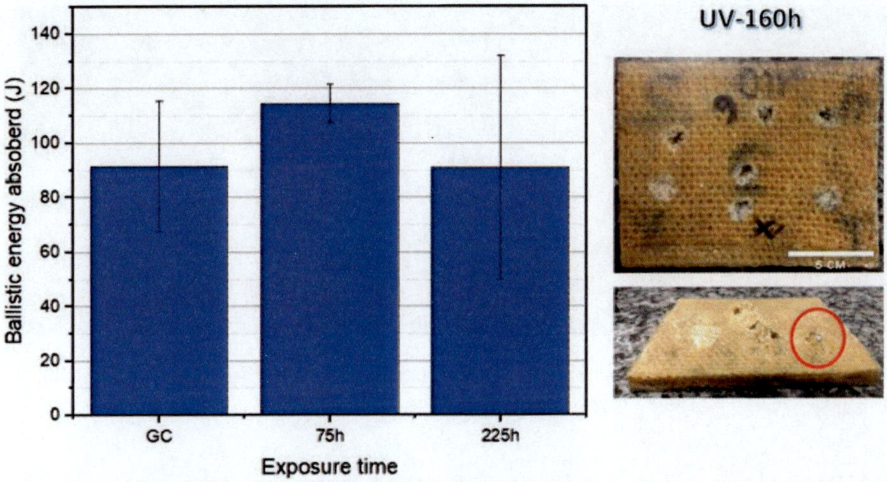

Fig. 3 Mean and standard deviation of the groups evaluated after ballistic test. (Color figure online)

perforation, as the projectile was captured by it, or the projectile had its path deviated, thus, not passing through the chronograph. An increase in absorption was noticed for the UV-225 h composite, this group obtained the absorption of 15% of the initial energy of the projectile. The FTIR spectra shown in Fig. 3 show the characteristic transmittance bands for the epoxy composite reinforced with sisal fabric.

Figure 4 shows the main transmittance bands observed in FTIR for sisal composites. The 3400 cm^{-1} band was referring to the OH bond, in addition, 2960 cm^{-1} is related to the connections of CH and CH3, 1622–1500 cm^{-1} referring to aromatic C–C and C=C aromatic isolated from lignin, 1260 cm^{-1}. Ether C–O–C (epoxide group)

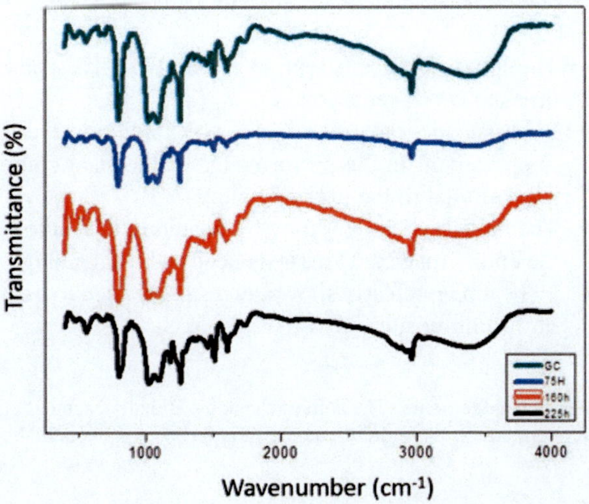

Fig. 4 FTIR spectrograms of the epoxy composite reinforced with sisal fabric before and after the times of 75, 160, and 225 h under UV radiation. (Color figure online)

Fig. 5 Epoxy composite reinforced with sisal fabric exposed to 225 h UV-B radiation after ballistic test equivalent to 0.22 caliber (Type I) [13]. (Color figure online)

and 790 cm^{-1} referring to the CH outside the plane. These analyses suggest that post-cure has occurred, providing an increase in crosslinks among polymer chains and reducing existing dipoles. The crosslinks corroborate to the composite became more resistant to the penetration of the projectile. Figure 4 shows the epoxy composite reinforced with sisal tissue exposed to 225 h of UV-B radiation after the ballistic test.

As can be seen in Fig. 5, the composite, after 225 h exposure to UV-B radiation and ballistic impact, performed well, despite the fact that the expected data, such as residual speed, were not obtained. Therefore, it is suggested that exposure to this radiation for the specified exposure times influenced the curing of the epoxy matrix, making it more resistant to projectile penetration.

Summary and Conclusions

- The epoxy composite reinforced with 30 vol% sisal fabric was exposed to different times to UV-B radiation.
- Materials degradations were expected; however, as for the ballistic test, an improvement in the performance of the sisal composite were observed directly proportional to the exposure time.
- The ballistic behavior may be related to the increase in the level of crosslinks in the epoxy matrix. More tests need to be carried out for this statement.
- FTIR did not clearly show any results related to treatment considered as post-cure of the composite.

Acknowledgements The authors thank the Brazilian agencies CAPES and CNPq for the financial support, and Prof. André Figueiredo for performing the ballistic tests.

References

1. Ozturk A, Kaynak C, Tincer T (2001) Effects of liquid rubber modification on the behavior of epoxy resin. Eur Polym J 37:2353–2363
2. Al-Turaif HA (2013) Surface morphology and chemistry of epoxy-based coatings after exposure to ultraviolet radiation. Prog Org Coat 76(4):677–681
3. Lu T, Solis-Ramos E, Yi Y, Kumosa M (2018) UV degradation model for polymers and polymer matrix composites. Polym Degrad Stab 154:203–210
4. Hang TTX, Dung NT, Truc TA, Duong NT, Van Truoc B, Vu PG, Olivier MG (2015) Effect of silane modified nano ZnO on UV degradation of polyurethane coatings. Prog Org Coat 79:68–74
5. Chee SS, Jawaid M, Sultan MTH, Alothman OY, Abdullah LC (2019) Accelerated weathering and soil burial effects on colour, biodegradability and thermal properties of bamboo/kenaf/epoxy hybrid composites. Polym Test 79:106054
6. Benzait Z, Trabzon L (2018) A review of recent research on materials used in polymer–matrix composites for body armor application. J Compos Mater 52(23):3241–3263
7. Martin AR, Martins MA, Mattoso LHC, Silva ORRF (2009) Caracterização química e estrutural de fibra de sisal da variedade Agave sisalana. Polímeros. 19(1):40–46
8. Rohen LA, Margem FM, Monteiro SN, Vieira CM, Fontes M, de Araujo B, Lima ES (2015) Ballistic efficiency of an individual epoxy composite reinforced with sisal fibers in multilayered armor. Mater Res 18(Suppl. 2):55–62
9. Sanjay MR, Madhu P, Jawaid M, Senthamaraikannan P, Senthil S, Pradeep S (2018) Characterization and properties of natural fiber polymer composites: a comprehensive review. J Clean Prod 172:566–581
10. Pires-de-Souza FDCP, Garcia LDFR, Hamida HM, Casemiro LA (2007) Color stability of composites subjected to accelerated aging after curing using either a halogen or a light emitting diode source. Braz Dental J 18(2):119–123
11. Mishra S, Sain M (2008) Long-term performance of natural-fiber composites. Prop Perform Nat-Fibre Compos 460–502
12. ASTM International (2000) G154–00 standard practice for operating fluorescent light apparatus for UV exposure of nonmetallic materials. ASTM International, West Conshohocken, PA
13. US Department of Justice, & National Institute of Justice (2000) NIJ 0101.04. Ballistic Resistance of Body Armor

Author Index

B. Li et al. (eds.), *Materials Engineering—From Ideas to Practice: An EPD Symposium in Honor of Jiann-Yang Hwang*, The Minerals, Metals & Materials Series,
https://doi.org/10.1007/978-3-030-65241-8

Subject Index

B. Li et al. (eds.), *Materials Engineering—From Ideas to Practice: An EPD Symposium in Honor of Jiann-Yang Hwang*, The Minerals, Metals & Materials Series,
https://doi.org/10.1007/978-3-030-65241-8